Status Survey and Conservation Action Plan

Eurasian Insectivores and Tree Shrews

Compiled by
R. David Stone
IUCN/SSC Insectivore, Tree Shrew and Elephant Shrew
Specialist Group

1995

Sultanate of Oman

Chicago Zoological Society

WWF

ENGLISH NATURE

Citation: IUCN. 1995. *Eurasian Insectivores and Tree Shrews – Status Survey and Conservation Action Plan.*
IUCN, Gland, Switzerland. 108 pp.

© 1995 International Union for Conservation of Nature and Natural Resources

Reproduction of this publication for educational and other non-commercial purposes is authorised without permission from the copyright holder, provided the source is cited and the copyright holder receives a copy of the reproduced material.

Reproduction for resale or other commercial purposes is prohibited without prior permission of the copyright holder.

ISBN 2-8317-0062-0

Available from:
IUCN Publications Services Unit, 219c Huntingdon Road, Cambridge, CB3 ODL, United Kingdom.
Tel: +44 1223 277894. Fax +44 1223 277175. E-mail: iucn-psu@wcmc.org.uk.
A catalogue of all IUCN publications can be obtained from the same address.

The designation of geographical entities in this book, and the presentation of the material, do not imply the expression of any opinion whatsoever on the part of IUCN concerning the legal status of any country, territory, or area, or of its authorities, or concerning the delimitation of its frontiers or boundaries.

Published by IUCN, Gland, Switzerland.

Camera-ready copy by the Nature Conservation Bureau Limited.

Printed by Information Press.

Paper: Huntsman Velvet 100gsm, wood fibre from sustainable forests, elementally chlorine-free.

Cover photo: Eurasian water shrew (*Neomys fodiens*). (Peter Vogel.)

Contents

	Page		Page
Foreword	iv	3.3 Species Accounts	60
		3.4 Captive Breeding of Tree Shrews	66
Executive Summary	v	3.5 Conservation Requirements of Tree Shrews – A Summary	66
Acknowledgements	vii		
1. Introduction	1	**4. Conservation Action Plan**	68
1.1 The Insectivora of Eurasia	1	4.1 Introduction	68
1.2 The Scandentia of South and South-east Asia	1	4.2 Conservation of Insectivores and Tree Shrews – Habitat Conservation	68
1.3 Geographical Coverage	2	4.3 Conservation of Insectivores and Tree Shrews – Field Surveys	70
1.4 The IUCN/SSC Insectivore, Tree Shrew and Elephant Shrew Specialist Group (ITSES)	4	4.4 Country Assessments of Status and Conservation Needs for Threatened Insectivores and Tree Shrews	70
1.5 Threats Facing the Eurasian Insectivores and Tree Shrews	4	4.5 Priorities for Action	85
1.6 Rationale and Objectives of the Action Plan	5		
2. The Insectivora of Eurasia	7	Appendix I Action Plan for the Russian Desman (*Desmana moschata*)	89
2.1 Introduction	7	Appendix II Action Plan for the Pyrenean Desman (*Galemys pyrenaicus*)	93
2.2 Taxonomic Classification	7	Appendix III IUCN Categories of Threat	95
2.3 Species Accounts	8	Appendix IV List of Threatened Insectivores and Tree Shrews	98
2.3.1 Family Erinaceidae: The Hedgehogs, Moonrats and Gymnures	8	Appendix V List of Members	100
2.3.2 Family Soricidae: The Shrews	17		
2.3.3 Family Talpidae: Moles and Desmans	46	**References**	103
3. The Scandentia of Asia	59		
3.1 Introduction	59		
3.2 Taxonomic Classification	60		

Foreword

One of the curiosities of eastern Nepal is a little-known insectivore known locally as *"pani musa"*, or "water rat". Knowing that its occurrence in the mountains to the east of Mt. Everest, on the border with Tibet, was still only suspected, I spent several weeks in 1973 seeking to confirm its occurrence there. With teams of local Sherpas, we trudged through many mountain torrents, turning over rocks, searching for evidence, and setting live traps. Our efforts were finally rewarded by capturing one individual of this elegant little water shrew, with amazingly silky fur, webbed feet with fringes, and a paddle-like tail. The local people were well aware of the existence of this animal, though they paid little attention to it because it was so innocuous and seemed to have so little to do with their affairs.

In this sense, the Nepalese were no different than most other people in the world: insectivores are basically unknown, unnoticed, and unloved. Yet as this Action Plan shows, these inconspicuous members of virtually all ecosystems throughout Eurasia are an important part of the ecological fabric of the region.

How, then, can greater public interest be aroused? Or at least, how can the conservation needs of these species be better incorporated in conservation plans? A useful first step might be to identify which on-going or planned projects are taking place in important habitats of key insectivores and tree shrews. For example, the Kerinci region of West Sumatra is the site of a major project receiving funding by the Global Environment Facility (GEF). This region supports populations of the critically-endangered Sumatran water shrew, the moonrat, the lesser gymnure, and no less than four species of tree shrew as well. Thus insectivores and tree shrews can help to enhance the importance of the Kerinci project for global biodiversity, thereby contributing to an important international effort.

Those interested in insectivores and tree shrews might then ask how projects already planned or under way might be modified to better address the needs of insectivores and tree shrews. Many of these projects include public information components, so those interested in these animals should devise approaches that will enable information about insectivores and tree shrews to contribute to such public information programmes. Several of these projects have research components, and these could be modified to incorporate appropriate research into tree shrews and insectivores.

Other important research questions for which answers might be sought could include:

- What role do insectivores play in maintaining the diversity of insect faunas?

- What role do moles and fossorial shrews play in the cycling of nutrients and water in forested ecosystems?

- How do tree shrews affect forest regeneration? Do they play any role in seed dispersal? Control insects which prey on seedlings?

- Given that some populations of widespread species of shrews are becoming isolated, can these populations be used for the study of speciation?

- What are the habitat requirements of key species, and what are the management implications of these requirements? If some species of tree shrews, for example, prefer early successional habitats, how should protected areas be managed to maintain such habitats?

The kinds of research implied by such questions will lead to an expanded appreciation of tree shrews and insectivores, demonstrating that they have an ecological importance far greater than is generally appreciated. I hope very much that this Action Plan leads quickly to new avenues of research that extend beyond basic taxonomy and distribution, and begin to investigate why insectivores and tree shrews are deserving of a significant investment by people in their conservation.

Jeffrey A. McNeely
Chief Biodiversity Officer
IUCN, Gland, Switzerland

Executive Summary

This Action Plan addresses the conservation needs of two unrelated groups of small mammals: the Insectivora (non-volant mammalian insectivores) and Scandentia (tree shrews) of Eurasia. It provides a review of what is currently known about the taxonomy, distribution, conservation status and requirements of almost 200 species. In assessing and evaluating these criteria, this Action Plan constitutes a first attempt to identify and prioritise species in terms of the degree of threat and need for conservation action. In addition, and through a series of specific recommendations, it seeks to stimulate further activities in order to promote a greater awareness of the extent of the threats facing certain species and the need for conservation action.

Insectivores are among the most numerous and widespread small mammals of the Eurasian region. For the purpose of this review, 180 species have been considered. The tree shrews form a separate, cohesive group of 19 species and are confined entirely to South and south-east Asia. Although unrelated, insectivores and tree shrews have a number of physical and behavioural features in common: both groups exhibit a number of primitive features, such as simplified dentition; all are predominantly insect-eating animals; and forests (both tropical and temperate) are one of the most important habitats of both groups. Insectivores, however, are not only confined to forests: many are fossorial, aquatic or terrestrial, living under a wide range of climatic and altitudinal conditions. They therefore face a much broader range of potential threats than tree shrews.

In spite of their ecological and evolutionary importance, insectivores and tree shrews remain some of the least well-known mammals. The majority of species are relatively unimportant to humans in economic terms; a limited number have been trapped for their fur or as a source of food, and only a few are considered pests. As most species are nocturnal, or secretive in other ways, many people are often unaware of the existence of these species. Tree shrews, perhaps the most obvious and visible of the two groups, are an exception but until recently have been almost totally ignored by field ecologists. In terms of their biomass, however, all of these species fill a major ecological role as predatory small mammals. Many also fill a range of unique niches. Yet, because these are not high profile animals, they are often overlooked in surveys when conservation or development plans are being drawn up for a region. For these reasons it is essential that the ecological requirements of these species be determined and a greater awareness generated of their conservation needs.

The most important single threat to insectivores and tree shrews is habitat destruction. This takes many forms: in western Europe, for example, removal of hedgerows, expanding road networks, intensive cultivation with widespread use of fertilizers and pesticides, and drainage of wetlands is having a noticeable impact on the numbers and distribution of many species. Water pollution, combined with the construction of hydro-electric dams and canals, threatens aquatic species throughout the region. Wetland drainage is also a major problem in south-east Asia. Of greater consequence in this region, however, is the level of human encroachment into tropical forests; slash-and-burn cultivation and logging are destroying vast areas of prime habitat for many species. Although little evidence is available on the effects such clearance may have on insectivore and tree shrew populations, it is highly likely that the combined effect of these processes could, in the long-term, lead to serious habitat fragmentation and isolation of small, vulnerable populations. Species at particular risk are those with already restricted distributions as well as the many monospecific genera covered in this Action Plan.

Perhaps the greatest problem in formulating conservation recommendations for insectivores and tree shrews is the current lack of data on the ecology, distribution and conservation status of the vast majority of species. The secretive nature of a great many of these animals has meant that vital information such as the extent of a species' range or population size is often lacking, or that few details exist on their natural history, ecology or conservation status. Using the most recent guidelines prepared by IUCN on how to identify threatened species, this report demonstrates that many species are endangered or threatened: 13 species have even been classified as "Critically Endangered" (Table 1). The current lack of reliable field data almost certainly prevents the identification of other potentially threatened species. It also impedes the formulation of definite conservation actions. On occasion, this problem is further compounded by uncertainties regarding taxonomic status. Such limitations hinder action that might help identify impending threats to a given species and its habitat, and enable remedial action to be taken. For these reasons, threats to insectivores and tree shrews often go unnoticed until it is too late to act. Future emphasis should be given to conducting additional field surveys to prepare and implement specific conservation programmes of these species and their habitats.

Despite the limitations imposed by the current lack of information on many species, this Action Plan still has an important role to fulfil in highlighting the need

Table 1. Critically endangered insectivores

Species	Country/Region
Hylomys parvus (dwarf gymnure)	West Sumatra (Indonesia)
Chimarrogale hantu (Malayan water shrew)	Malay Peninsula
Chimarrogale sumatrana (Sumatra water shrew)	West Sumatra (Indonesia)
Sorex kozlovi (Kozlov's shrew)	Tibet
Sorex cansulus (Gansu shrew)	Gansu Province, China
Soriculus salenskii (Salenski's shrew)	Sichuan Province, China
Crocidura dhofarensis	Oman
Crocidura jenkinsii	South Andaman Island (India)
Crocidura negrina	Negros Island, the Philippines
Suncus ater (black shrew)	Sabah (Malaysia)
Suncus mertensi	Flores Island, Indonesia
Euroscaptor parvidens	Vietnam
Talpa streeti (Persian mole)	North-west Iran

for action for at least those species which have been identified as threatened. In reviewing this information, a number of priority actions have been identified for the conservation of insectivores and tree shrews, specifically to draw attention to the need for, and to promote:

- further action – particularly field surveys of poorly-known species – in order to assess distribution and conservation status and to identify remedial action where appropriate;

- contact with field ecologists, taxonomists and geneticists, who are asked to communicate any observations on insectivores and tree shrews in Eurasia to help improve the scant information on those rarer, poorly known species;

- cooperation with ongoing conservation efforts throughout Eurasia so that, when appropriate, recommended and approved actions by ITSES – the Insectivore, Tree Shrew and Elephant Shrew Specialist Group – can be incorporated within existing or intended projects;

- the interest of universities, associated institutes and expeditions, which are often of considerable importance by stimulating interest and research related to conservation; and

- public awareness of the importance of habitat conservation; what benefits a shrew, for example, will almost certainly be of great benefit to a much wider range of wildlife. Often, this may also result in positive fringe benefits for local human communities.

In the first instance, the target audience of this Action Plan is ecologists and researchers. Only when more specific details are known about the conservation status and threats to endangered species can action be pursued at other levels. In publishing this Action Plan an appeal is therefore made to scientists, members of other SSC groups and the wider public to gather and promote more information on Eurasian insectivores and tree shrews. It is anticipated that this will help stimulate further action and enable ITSES to prepare a more specific list of priority conservation projects for Eurasian Insectivora and Scandentia, which it can use as a definite statement when applying to relevant authorities for further action.

Use of this Action Plan can also serve objectives additional to those listed above, for example, assisting in regional assessments of biological diversity and habitats. These could vary in scope and nature, depending on the specific issues and/or projects being dealt with. It is ITSES' hope that this Action Plan will reach a wide audience and will help stimulate and direct future field projects on behalf of the conservation of insectivores and tree shrews in the Eurasian region.

Acknowledgements

The compilation of this Action Plan could not have been achieved without the assistance of the members of ITSES – the Insectivore, Tree Shrew and Elephant Shrew Specialist Group, for it was they who provided much valuable information at the outset, as well as constructive comments on various data sheets through the progression of the document. Acknowledgement is also made to the many researchers in insectivore taxonomy and ecology whose efforts throughout the years have contributed greatly to our present understanding of insectivore ecology, and whose works are reflected in this presentation.

Particular thanks are extended to Dr. Robert Hoffmann and Dr. Gordon Corbet for their thoughtful comments and considerable patience in helping the author to unravel many of the mysteries surrounding insectivore classification. Their efforts over the years, together with those of many other experts, have helped to clarify the mystery of small mammal taxonomy and has meant that this work could become a reality and not a dream.

Sincere thanks are also expressed to the following for their input and constructive comments to early versions of this manuscript: Hisashi Abe, Sara Churchfield, Mariano Gimenez-Dixon, Linette Humphrey, Rainer Hutterer, Jeffrey A McNeely, Martin Nicoll, Junaidi Payne, Manuel Ruedi, Peter Vogel, and Terry Yates.

The unwavering support and patience of Simon Stuart and Martin Nicoll has played a greater role in the finalisation of this Action Plan than either could possibly realise.

The support of WWF – World Wide Fund For Nature and the IUCN/SSC Sir Peter Scott Action Plan Fund (established through a donation from the Sultanate of Oman) provided support for the production of this document. This assistance is gratefully acknowledged.

Chapter 1

Introduction

With 423 species recognised worldwide (Hutterer, 1993), insectivores (Order Insectivora) are the third largest order of mammals after rodents and bats. They are largely confined to the northern temperate zones of North America, Europe, the former Soviet Union, Africa, and southern Asia. The tree shrews (Order Scandentia) constitute a much smaller group of just 19 species, all of which are restricted to South and south-east Asia; it is the best single group for defining the Indo-Malayan Realm. Another group of small, terrestrial mammals which share many of the primitive features of Insectivora and Scandentia is the Order Macroscelidea, the elephant shrews of Africa. Fifteen species are recognised in this order (Wilson and Reeder, 1993).

Apart from their small size, members of these three groups are characterised by their general tendency to feed on invertebrates, although many species also take a range of small amphibians, seeds and fruit. Most are opportunistic hunters. To assist with feeding, many species have developed additional specialisations, such as the paralysing bite of the mole or water shrew, as well as prey-caching behaviour among many terrestrial and fossorial species.

Most insectivores are highly secretive species and often nocturnal. Their habitats range from arid and semi-arid conditions to montane forests, and from semi-aquatic tendencies to living under desert and Arctic conditions. The majority are terrestrial, but some are fossorial, semi-fossorial, aquatic or even arboreal in habit. Combined with a small body size, their secretive nature has resulted in the majority of these species being neglected or overlooked in many field surveys. As a result, little is known about the distribution, behaviour or ecology of the majority of these animals.

This Action Plan has been prepared as a first step to addressing this oversight. Prepared by the IUCN/SSC Insectivore, Tree Shrew and Elephant Shrew Specialist Group (ITSES), it examines the current status and conservation needs of the Insectivora and Scandentia of Eurasia. Similar coverage has already been compiled for the Insectivora and Macroscelidea of Africa (Nicoll and Rathbun, 1990).

1.1 The Insectivora of Eurasia

The Order Insectivora is ancient. Fossil evidence now indicates that the most primitive placental mammals were insectivores. Today's species are considered representative of the ancestral stock from which modern mammals are derived. Their descendants have retained a number of primitive features, including a non-specialised dental pattern, a simple brain, a reliance on their sense of smell (as opposed to vision), unspecialised limbs and a generalised quadrupedal mode of locomotion.

The Insectivora comprise a wide variety of largely insectivorous mammals which have often been grouped together because of common morphological traits rather than a clear, recent common ancestry. The taxonomic classification of insectivores has been reviewed in detail by Cabrera (1925), van Valen (1967), Gureev (1979), Butler (1988), Yates (1984) and Hutterer (1993). As a result of these, and other, investigations the Family Tupaiidae (tree shrews), which was once included in the Order Insectivora, has now been assigned a separate order – Scandentia. Likewise, elephant shrews (Family Macroscelididae) are now placed in a distinct order – Macroscelidea.

As a result of these reviews, the Order Insectivora is now seen to contain six extant families, of which just three are represented in the Eurasian region: Erinaceidae (hedgehogs), Soricidae (shrews) and Talpidae (moles and desmans) (see Table 1.1).

1.2 The Scandentia of South and South-east Asia

Tree shrews (Order Scandentia) are small mammals found in the tropical rainforests of southern and south-eastern Asia. Superficially tree shrews resemble small tree squirrels – the genus name *Tupaia* is derived from the Malay *'Tupai'*, meaning 'squirrel-like animal' – but it is now recognised that the two groups differ widely in anatomy and behaviour. Most species of tree shrew are

Table 1.1. Taxonomic summary of the Eurasian Insectivora (Hutterer, 1993)	
Genus	Number of species
Family Erinaceidae (hedgehogs)	
Atelerix	1
Echinosorex	1
Erinaceus	3
Hemiechinus	6
Hylomys	4
Mesechinus	2
Podogymnura	2
Subtotal	19
Family Soricidae (shrews)	
Anourosorex	1
Blarinella	2
Chimarrogale	6
Crocidura	55
Diplomesodon	1
Feroculus	1
Nectogale	1
Neomys	3
Solisorex	1
Sorex	34
Soriculus	10
Suncus	11
Subtotal	126
Family Talpidae (moles)	
Desmana	1
Euroscaptor	6
Galemys	1
Mogera	7
Nesocaptor	1
Parascaptor	1
Scapanulus	1
Scaptochirus	1
Scaptonyx	1
Talpa	9
Uropsilus	4
Urotrichus	2
Subtotal	35
Total	**180**

Table 1.2. Taxonomic summary of the Asian tree shrews (Wilson, 1993)	
Genus	Number of species
Family Tupaiidae	
Anathana	1
Dendrogale	2
Ptilocercus	1
Tupaia	14
Urogale	1
Total	**19**

semi-terrestrial, but all are agile climbers; tree squirrels are almost exclusively arboreal.

Tree shrews form a cohesive group of 19 species (Table 1.2) confined entirely to South and south-east Asia. None of the five genera covers the entire geographical range of the order; the genus *Tupaia* is by far the most widespread of these. The greatest number of species is found on the island of Borneo. This concentration is believed, in part, to be a consequence of the great size of the island and the resulting wide range of available habitats. But it is also possible that Borneo was the centre from which the adaptive radiation of modern tree shrew species began (Martin, 1984). Borneo was also joined to the Asian land mass for longer periods of time during the most recent Ice Age, enabling more species to migrate to it than to present-day offshore islands, such as many of those of the Indonesian archipelago.

1.3 Geographical Coverage

The Eurasian land mass, spreading halfway around the globe, comprises the world's largest land mass. This region covers two major biogeographical realms – the Palaearctic and Indomalayan. Included in this vast region are a wide range of habitats, ranging from tundra to tropical rainforest and including coniferous and deciduous forests, prairie and dry steppes, Mediterranean scrub vegetation and desert conditions.

For the purpose of this Action Plan the Eurasian region is seen to extend from the Great Blasket Island, West Ireland (10°32'W) to the Diomede islands in the Bering Straits (169°E), and from latitude 80°N (well inside the Arctic Circle) to Pamana on the island of Roti (Indonesia) in the Timor Sea (10°59'S) (see Map 1.1). Selected countries of North Africa (Morocco, Algeria, Tunisia, Libya and Egypt) are also included in this review as the range of a number of threatened shrew species extends across both North Africa and Eurasia.

In addition to the extensive land mass considered, a large number of islands are covered in this review. This includes all of the islands in the Mediterranean Sea, Socotra (Yemen) in the Gulf of Aden, the Andaman and Nicobar Islands (India), all of the islands in the Indonesian archipelago (including New Guinea), the Philippines, Hainan (China), Hong Kong (UK), Taiwan, Japan, Sakhalin Island (Russia) in the sea of Okhotsk, and the

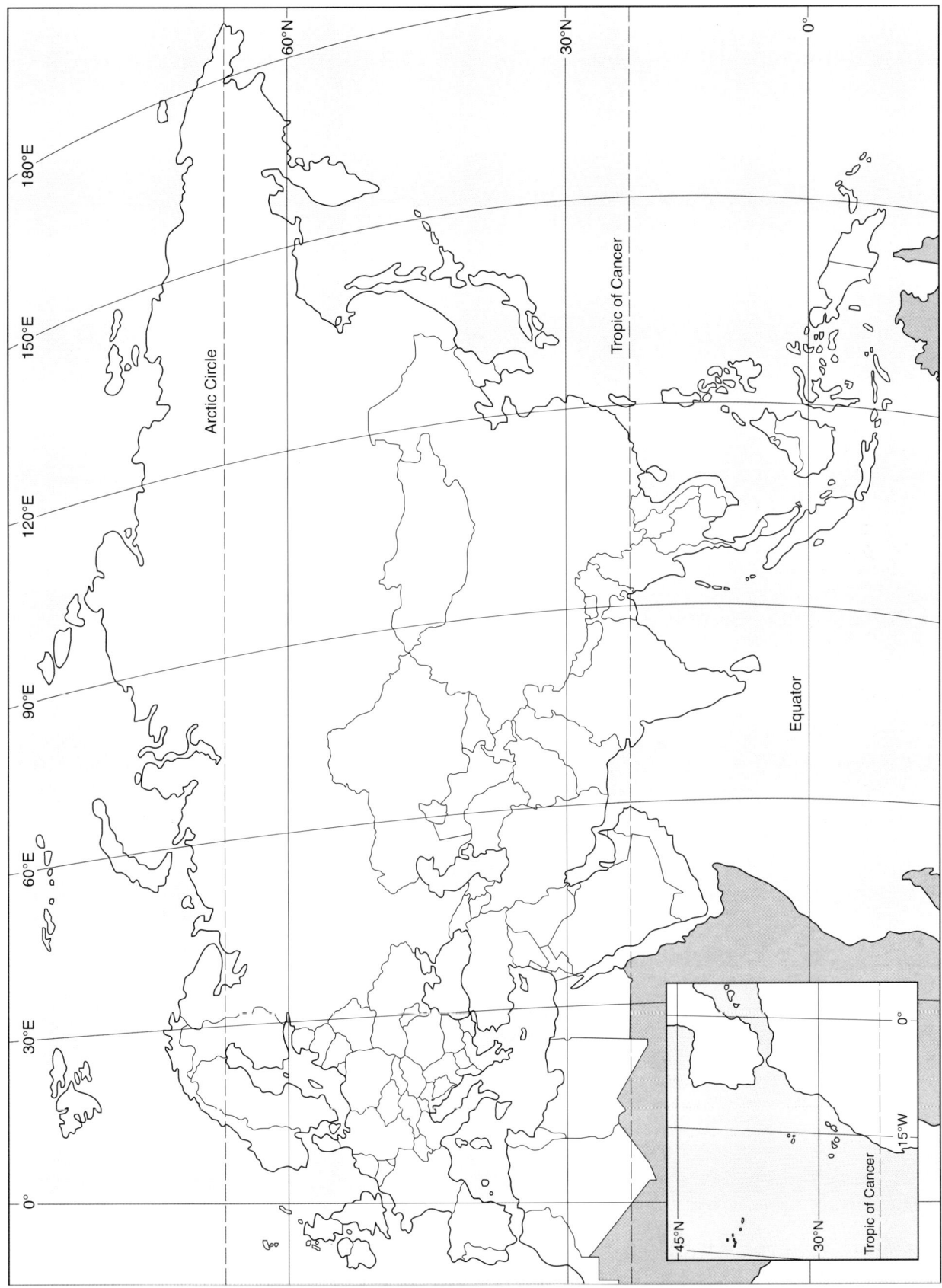

Map 1.1. Region covered by this publication, showing Eurasian and North African countries.

Kuril Islands (Russia). Although lying outside of the geographical limits outlined above, the Canary Islands (Spain), Madeira (Spain) and the Azores (Portugal) are also included within this review. Excluded are the islands of the Pacific Ocean, the Seychelles, Maldives and Chagos Island (UK).

1.4 The IUCN/SSC Insectivore, Tree Shrew and Elephant Shrew Specialist Group (ITSES)

The IUCN Species Survival Commission now has 104 Specialist Groups covering a wide range of taxonomic groups and conservation techniques worldwide. The common objective of these groups is to provide technical advice and guidance to the world conservation community (usually taking the form of governments and member organizations of IUCN) on the conservation of the species within their brief. One of the main activities of all taxa-focused Specialist Groups is to prepare Action Plans which outline the conservation priorities for their species.

This Action Plan covering Eurasia is the second in a series of regional Action Plans being prepared by ITSES. An Action Plan for the Conservation of African Insectivora and Elephant Shrews was published in 1990 (Nicoll and Rathbun, 1990). A third action plan is currently in preparation for the Insectivora of the Americas.

Although the Species Survival Commission carried out some work on the Insectivora during the early 1980s, the Specialist Group in its current form was re-established in 1986. Its brief was also extended at the time to include the Scandentia and Macroscelidea. A list of current members is included as Appendix V.

In addition to providing a forum for expert review and discussion, a crucial role for ITSES is in the development and promotion of conservation projects, as well as the identification of potential funding sources. In this role, ITSES has already had some success in promoting conservation action for a number of species as well as increasing public and scientific awareness for the conservation needs of species within its brief. Wherever possible these projects will be fully integrated within other larger and more broad-based habitat protection programmes.

One of the major difficulties encountered in promoting the need for additional actions is that few insectivores, tree shrews or elephant shrews are of direct economic or even aesthetic value for mankind. Yet it is recognised that these are not 'insignificant' species, since they play a critical role in maintaining ecological diversity – not only in the role they play as key predators of invertebrates and as potential prey to a wide range of other species, but also in terms of their genetic diversity and the wide range of habitats and niches to which they have adapted. As research continues to unfold many fascinating aspects of these species' ecology it is essential that the habitat requirements and conservation status of these species is also better realised. Only then can measures be taken to identify and protect any threatened species.

In addition to its field and awareness activities, the Group also produces an occasional newsletter which aims to share information among the researchers working on the conservation of these poorly-known species. The Group now includes some 34 members, among them small mammal ecologists, scientists working in universities, zoological gardens and natural history museums, as well as interested amateur naturalists. It is not the Group's intention to be an exclusively scientific-based community; membership is open to anyone with a commitment to the conservation of those species which fall within the brief of the Group. Additional information on ITSES may be obtained by writing to the Chairman, or directly to the SSC Secretariat in Gland.

1.5 Threats Facing Eurasian Insectivores and Tree Shrews

Unlike the majority of other mammalian groups few insectivores or tree shrews are subject to direct human interference through persecution. Although a few species of insectivores were once hunted for their valuable and highly prized fur (principally the European mole *Talpa europaea* and the Russian Desman *Desmana moschata*), this practise is no longer widespread and does not represent a threat to extant species. European moles are still considered a pest species in some countries, but this tends to apply only to regions of highly developed agriculture (see Stone, 1989 for overview).

A highly specialised species, the threatened Russian desman (*Desmana moschata*), shown here with the late Gerald Durrell, represents the plight of all semi-aquatic insectivores in the Eurasian region. (Photo by John Hartley)

The most important single reason for the decline of whole species and genera is probably habitat destruction. This takes many forms in Eurasia, each affecting a wide range of species. In western Europe, for example, removal of hedgerows, an ever-expanding road network, increasing cultivation with large scale additions of fertilizer and pesticides, and drainage of wetlands is having a noticeable impact on the number and distribution of many species. Increasing pollution of freshwater resources, combined with the construction of hydro-electric dams and canals play havoc on aquatic species such as the desmans and water shrews.

In south-east Asia, human encroachment into tropical forests either as a result of logging or slash-and-burn cultivation is destroying prime habitat for many species, including a large number of insectivores and tree shrews. Wetland drainage is also a major continuing problem in south-east Asia. In the long-term, such processes lead to habitat fragmentation and isolation of small, vulnerable populations.

Finally, for the vast majority of species under consideration in this Action Plan, one of the most worrying points is how little is actually known about the great majority of species. For these, the lack of data on basic ecology, distribution and conservation status is a major concern since it makes it more difficult to identify actions that might help detect impending threats to a given species and its habitat, and enable remedial action to be taken.

1.6 Rationale and Objectives of the Action Plan

Insectivores and tree shrews are not considered among the most charismatic animals. Many people are not even aware of their existence. For these reasons, threats to insectivores and tree shrews often go unnoticed until it is too late to act. The present Action Plan for Eurasian Insectivora and Scandentia has been compiled to address this oversight.

In presenting this information, it is recognised that the Action Plan has many limitations. The secretive and often nocturnal nature of a great many of the species covered in this report has meant that few details are known on the natural history, ecology or conservation status of the majority of species. However, within the limits imposed by the current lack of information, it is firmly believed that this Action Plan has an important role to fulfil. Previous experience with the publication of such Action Plans supports this: the 1986 IUCN Red List of Threatened Animals (IUCN, 1986), for example, included just eight species of Insectivora. Following the publication of the Action Plan for African Insectivora and Elephant Shrews, 23 new species were added to this list when published again in 1990 (IUCN, 1990). More important, many worthwhile field programmes and related activities have arisen from this exercise, demonstrating the value of such an activity.

One of the principal objectives of this Action Plan, therefore, is to assess the current status of the Eurasian Insectivora and Scandentia. Within its brief, there are some 200 species; "satisfactory" information is thought to exist for maybe just a dozen of these. For many, a single museum specimen represents our total knowledge of the species. At the outset of this exercise, just five species were recognised as threatened in some way: *Podogymnura truei*, *Crocidura tenuis*, *C. zimmermanni*, *Desmana moschata* and *Galemys pyrenaicus*. In publishing this Action Plan, which applies a new series of criteria for defining a species' conservation status (IUCN, 1995), a total of 68 species has been identified as being in need of some form of protection, while another six require further research in order to assess their status (see Chapter 4 for further details).

In addressing these issues, this Action Plan provides a concise review of what is known about the taxonomy, distribution and conservation status of all members of the Insectivora and Tupaiidae in Eurasia. Information on taxonomy, distribution and recommendations has been obtained from a wide range of published sources (see References), unpublished reports and by direct communications with members of ITSES. To assist and encourage further investigations of these species, a brief description of each is included, together with pertinent notes on behaviour and ecology which might assist future researchers in locating and identifying these species.

In publishing this account – and recognising its many weaknesses at the present time – an appeal is made to scientists, members of other SSC groups and the wider public for more information on Eurasian insectivores and tree shrews.

In summary, the principal objectives of this Action Plan are to:

- summarise information on the taxonomic status, distribution, and conservation status and requirements of Eurasian Insectivora and Scandentia;

- prioritise species in terms of the need for conservation action;

- prepare the groundwork for the preparation of a list of priority conservation projects for Eurasian Insectivora and Scandentia (see, for example, Appendix I and II);

- stimulate further action – particularly field surveys of poorly-known species – in order to assess distribution and conservation status and to identify remedial action where appropriate;

- promote and maintain contact with field ecologists, taxonomists and geneticists who are asked to communicate any observations on insectivores and tree shrews in Eurasia to help improve the scant information on those rarer, poorly known species;

- liaise and cooperate with ongoing conservation efforts throughout Eurasia so that, when appropriate, recommended and approved actions by ITSES can be incorporated within existing or intended projects;

- promote and encourage universities, associated institutes and expeditions, which are often of considerable importance by stimulating interest and research related to conservation; and

- create and increase public awareness of the importance of habitat conservation; what benefits a shrew, for example, will almost certainly be of great benefit to a much wider range of wildlife. Often this may also result in positive fringe benefits for local human communities.

Chapter 2

The Insectivora of Eurasia

2.1 Introduction

Insectivores are among the most numerous and widespread small mammals of the Eurasian region. Their abundance, diversity and ecological role as key small predators has long been recognised by ecologists, and been the subject of a large number of field projects throughout the region. Many research efforts have, however, been hampered by inadequate means of detecting, observing or following these small animals in their natural environment. Recent advances such as the application of miniature radio transmitters should help overcome many of these problems and it is anticipated that much valuable ecological data will be forthcoming as a result of increased usage of these techniques.

A second problem which is frequently encountered when sifting through literature on these species is taxonomic confusion. Many species and subspecies still await investigation and confirmation of taxonomic affiliations. Once again, however, recent advances and applications of various biochemical analyses have helped unravel many of the greyer areas of classification. It is expected that these applications, when even more widely applied, will result in further changes to the taxonomy of many species.

For these reasons, it is important to adopt an approach for this Action Plan that is representative of the current status of each species, but which also reflects the growing efforts that are under way to untangle the taxonomic confusion of many groups. This is especially the case when dealing with the Family Soricidae (shrews). Every effort has been made to ensure that the following information is as up-to-date as possible; however, further studies, such as those described above, will inevitably result in some changes to the following accounts.

2.2 Taxonomic Classification

The taxonomic classification of the Insectivora has been turbulent and confusing for a considerable number of years. At one stage the Order served as a convenient "dumping ground" for other primitive forms of insect-eating mammals. A common arrangement at one time was to divide the Insectivora into two suborders: the Lipotyphla (with the living families Erinaceidae, Solenodontidae, Nesophontidae, Tenrecidae, Chrysochloridae, Soricidae, and Talpidae), and the Menotyphla, with the families Macroscelididae and Tupaiidae.

Van Valen (1967) considered the Order Insectivora to be restricted to the families Tupaiidae, Macroscelididae, Cynocephalidae (now in the order Dermoptera), Erinaceidae, Talpidae, Nesophontidae, and Soricidae. The families Tenrecidae, Solenodontidae, and Chrysochloridae, were placed in a separate order – Deltatheridia. Eisenberg (1981) also united the Tenrecidae and Chrysochloridae in a separate order, the Tenrecomorpha.

In a more recent review, McKenna (1975) elevated the Insectivora to the rank of a grandorder containing two orders: the Erinaceomorpha, with the family Erinaceidae; and the Soricomorpha, with the living families Talpidae, Nesophontidae, Soricidae, Solenodontidae, Tenrecidae, and Chrysochloridae. The Macroscelididae were placed in a separate order – the Macroscelidea – which, together with the order Lagomorpha, was put in the grandorder Anagalidia. Still another grandorder, the Archonta was considered to include the orders Scandentia (with the family Tupaiidae), Dermoptera, Chiroptera, and Primates. A modified version of this arrangement was suggested by Novacek (1986), who ranked the Insectivora as a superorder, with the Lipotyphla being the only living order, containing two suborders: the Erinaceomorpha for the Erinaceidae, and the Soricomorpha for the superfamilies Tenrecoidea and Soricoidea (Walker, 1991).

A consensus of current thought on these matters, as expressed by Yates (1984), was that the Insectivora should still be regarded as an order but one that contained only the Lipotyphlan families (including the Erinaceidae) and that the Tupaiidae and Macroscelididae are not closely related to this group and should be placed in separate orders. This broad formula is adopted for the current Action Plan. On a finer level, the recent taxonomic reviews by Hutterer (1993) and Wilson (1993) are followed in this document to describe the nomenclature for Insectivora and Scandentia, respectively.

2.3 Species Accounts

In this section, details are presented on insectivore species in Eurasia. Coverage includes 19 species of hedgehog, 126 shrew species and 35 species of mole. Standard species accounts are given for each species. These include notes on the species' taxonomic nomenclature, a brief physical description of each species, together with notes on its distribution, habitat preferences and ecology, where known. The "IUCN Category of Threat" is also given, using the latest categories and criteria adopted in 1995 (see Box 2.1 and Appendix III).

2.3.1 Family Erinaceidae: The Hedgehogs, Moonrats and Gymnures

This family of seven genera and 19 species (Table 2.1) is widely dispersed throughout Africa, Europe and Asia, ranging as far north as the limits of the northern deciduous forests, south to the Sahara desert and east towards, and including, Tioman Island (Peninsular Malaysia), Sumatra and Java (Indonesia), Borneo and Mindanao Island in the Philippines archipelago (Frost *et al.*, 1991).

The family Erinaceidae consists of two sub-families: the Erinaceinae – four genera of hedgehogs – and the Hylomyinae (formerly Echinosoricinae) – three genera of gymnures. The Erinaceinae are a widely distributed group throughout Eurasia and are readily distinguished by the barbless spines (adults have an average of 5000) on their back and sides. The remainder of the body – face, limbs and underparts – is covered in coarse hair. In contrast, members of the Hylominae are entirely Asian in distribution and are characterised by their lack of spines.

Members of this family vary considerably in size and appearance. Body length (head and body only), for example, ranges from 105–445mm, while tail length may vary from 10–325mm. The south-east Asian moonrat or gymnure *Echinosorex*, is the largest species, while *Hylomys*, the lesser gymnure, another Asian species, is the smallest. The snout of gymnures and hedgehogs is elongate and blunt and the eyes and ears are well developed. The tail is usually hairy. With the exception of the spines and their associated musculature, the body plan of hedgehogs and gymnures is very primitive.

The habitats of hedgehogs and gymnures include forested and bushy areas, steppes and deserts. Almost all species are nocturnal, sheltering during daylight under logs, in rock crevices or in shallow burrows. Species from the temperate region always construct a nest of dried grasses and leaves. This is particularly important for the European hedgehog (*Erinaceus europaeus*) which hibernates during the northern winter months.

Like other insectivores, hedgehogs and moonrats eat a variety of prey. European hedgehogs feed primarily on invertebrates including earthworms, slugs, beetles and caterpillars, while the Daurian hedgehog (*Mesechinus dauuricus*) of the Gobi Desert has been reported to feed mainly on small rodents. Moonrats frequently search for prey in water, feeding on a range of crustaceans, molluscs and even fish. Most species will also feed on fruit and berries.

Despite their large size (relative to other insectivores), hedgehogs and gymnures have not been well studied. The sole exception to this is perhaps the European hedgehog (*E. europaeus*) which has been the subject of many detailed laboratory and field studies. However, little is still known about the distribution, ecology or conservation status of the majority of Asian species.

Sub-family Erinaceinae

GENUS *ATELERIX*

Four species are recognised in this genus (Meester *et al.*, 1986) of which just one *A. algirus* is represented in Eurasia. The remaining species are exclusively African.

Algerian hedgehog (*Atelerix algirus*)

Taxonomy: *Atelerix algirus* Lereboullet 1842.

IUCN Category of Threat: Lower Risk (subcategory Least Concern).

Description: The Algerian hedgehog is somewhat smaller that the European species, measuring some 200–250mm. In appearance, the Algerian hedgehog is noticeably paler in colour than most examples of the European hedgehog, although *Erinaceus europaeus* may be quite pale in colour in southern Spain. The most reliable means of distinguishing *A. algirus* is the spine-free 'parting' on the crown of the head, which is wider than in other species.

Distribution: Although this species is widespread in north-western Africa (ranging from Western Sahara and Morocco to Libya), its Eurasian distribution is confined to south-west Europe, particularly on Malta (Malec and Storch, 1972), the Balearic Islands and the Mediterranean coastline of Spain and parts of south-eastern France,

Box 2.1. IUCN Categories of Threat

Extinct (EX)

A taxon is Extinct when there is no reasonable doubt that the last individual has died.

Extinct in the Wild (EW)

A taxon is Extinct in the Wild when it is known only to survive in cultivation, in captivity or as a naturalised population (or populations) well outside the past range. A taxon is presumed extinct in the wild when exhaustive surveys in known and/or expected habitat, at appropriate times (diurnal, seasonal, annual), throughout its historic range have failed to record an individual. Surveys should be over a time frame appropriate to the taxon's life cycle and life form.

Critically Endangered (CR)

A taxon is Critically Endangered when it is facing an extremely high risk of extinction in the wild in the immediate future, as defined by any of the criteria A to E outlined in Appendix III.

Endangered (EN)

A taxon is Endangered when it is not Critically Endangered but is facing a very high risk of extinction in the wild in the near future, as defined by any of the criteria A to E (see Appendix III).

Vulnerable (VU)

A taxon is Vulnerable when it is not Critically Endangered or Endangered but is facing a high risk of extinction in the wild in the medium-term future, as defined by any of the criteria A to D (Appendix III).

Lower Risk (LR)

A taxon is Lower Risk when it has been evaluated and does not satisfy the criteria for any of the categories Critically Endangered, Endangered or Vulnerable. Taxa included in the Lower Risk category can be separated into three sub-categories:

1. Conservation Dependent (cd). Taxa which are the focus of a continuing taxon-specific or habitat-specific conservation programme targeted towards the taxon in question, the cessation of which would result in the taxon qualifying for one of the threatened categories above within a period of five years;

2. Near Threatened (nt). Taxa which do not qualify for Conservation Dependent, but which are close to qualifying for Vulnerable; and

3. Least Concern (lc). Taxa which do not qualify for Conservation Dependent or Near Threatened.

Data Deficient (DD)

A taxon is Data Deficient when there is inadequate information to make a direct, or indirect, assessment of its risk of extinction based on its distribution and/or population status. A taxon in this category may be well studied, and its biology well known, but appropriate data on abundance and/or distribution is lacking. Data Deficient is therefore not a category of threat or Lower Risk. Listing of taxa in this category indicates that more information is required and acknowledges the possibility that future research will show that threatened classification is appropriate. It is important to make positive use of whatever data are available. In many cases great care should be exercised in choosing between DD and threatened status. If the range of a taxon is suspected to be relatively circumscribed, if a considerable period of time has elapsed since the last record of the taxon, threatened status may well be justified.

Not Evaluated (NE)

A taxon is Not Evaluated when it is has not yet been assessed against the criteria.

Source: IUCN (1995)

Table 2.1. Classification of the Erinaceidae[1]	
Genus	**Species**
Sub-family Erinaceinae	
Atelerix	A. algirus[2]
Erinaceus	E. amurensis
	E. concolor
	E. europaeus
Hemiechinus (Hemiechinus)	(H.) auritus
	(H.) collaris
Hemiechinus (Paraechinus)	H. (P.) aethiopicus
	H. (P.) hypomelas
	H. (P.) micropus
	H. (P.) nudiventris
Mesechinus	M. dauuricus
	M. hughi
Sub-family Hylominae	
Echinosorex	E. gymnura
Hylomys	H. hainanensis
	H. parvus
	H. sinensis
	H. suillus
Podogymnura	P. aureospinula
	P. truei

[1] The latest revision (Frost et al., 1991) recognises Mesechinus as a distinct genus and Paraechinus and Hemiechinus as subgenera of the latter.
[2] The sole representative of this African genus (four species) in Eurasia.

where it was probably introduced. It has also been introduced on Fuerteventura and Tenerife, the Canary Islands (Niethammer, 1972). An isolated population exists in mid-western France where it has almost certainly been introduced.

Habitat: The preferred habitat of this species has not been well defined.

Ecology and behaviour: Almost nothing is known about this species. Unlike *E. europaeus* it does not appear to hibernate during the winter months.

GENUS *ERINACEUS*

The genus *Erinaceus* contains three species. The range of the genus is very wide, extending from the British Isles across Europe and Russia to Manchuria, Korea and northern and eastern China. In Asia Minor its range extends east to Transcaucasia and Iran. *E. europaeus* has also been introduced to New Zealand.

Manchurian (Amur) hedgehog (*Erinaceus amurensis*)

Taxonomy: *Erinaceus europaeus* var. *amurensis* Schrenk 1859.

IUCN Category of Threat: Lower Risk (subcategory Least Concern).

Description: This species is easily distinguished by its thick body, the back of which is covered with long sharp spines up to 24mm long, and pale yellow in appearance. Other parts of the body are covered with fur composed of coarse hair. Body length is approximately 277mm.

Distribution: This species occurs in lowland China from about 29°N (i.e. a little south of the Yangtze) north to the Amur Basin and Korea. It may also occur in Sichuan, China (Corbet, 1992).

Habitat: Forest and grassland.

Ecology and behaviour: The ecology of this species is poorly known.

Eastern European hedgehog (*Erinaceus concolor*)

Taxonomy: *Erinaceus concolor* Martin 1838.

IUCN Category of Threat: Lower Risk (subcategory Least Concern).

Description: Similar in size and general appearance to *E. europaeus*. It may be distinguished by the presence of a distinctive white breast which contrasts with the dark-coloured belly.

Distribution: This species is widely distributed throughout eastern Europe, overlapping with *E. europaeus* in a zone from western Poland to the Adriatic. *E. concolor* is the only species of hedgehog on Crete and some other Greek islands. Its range also extends to Israel and Iran.

Habitat: Similar to *E. europaeus*.

Ecology and behaviour: This species has not been studied in any detail and little is known of its ecology. It is probably similar to *E. europaeus* in behaviour.

Western European hedgehog (*Erinaceus europaeus*)

Taxonomy: *Erinaceus europaeus* Linnaeus 1758.

IUCN Category of Threat: Lower Risk (subcategory Least Concern).

Description: The European hedgehog measures 225–275mm in length and weighs about 400–1100g. It may be distinguished from the eastern hedgehog (*E. concolor*) by the absence of a white breast patch in the western form. In southern Spain, *E. europaeus* is very pale in colour, with many wholly white spines and may be distinguished from the very similar Algerian hedgehog (*Atelerix algirus*) by the very narrow spine-free parting on the crown.

Distribution: This species is widely distributed throughout most of western Europe, including the British Isles. The zone of overlap with the Eastern hedgehog runs from the Baltic to the Adriatic and is about 200km wide in former Czechoslovakia. The range of *E. europaeus* expanded northward in Scandinavia during the 20th century (Kristiansson, 1981). It is also present in northern Russia and western Siberia. This species has been deliberately introduced to New Zealand.

Habitat: *E. europaeus* is found in woodland wherever there is ground vegetation, but is also abundant in grassland, especially when adjacent to woodland, hedgerow or scrub. In the Alps it may be found up to an altitude of 2000m in the dwarf pine zone, but will not live above the tree line.

Ecology and behaviour: Hedgehogs live on the surface of the ground without burrowing or climbing to any extent. Predominantly nocturnal, they are occasionally active during daylight, particularly during autumn. Undergoes hibernation during the winter months in the northern hemisphere. Details of its ecology may be found in Morris (1983).

GENUS *HEMIECHINUS*

The status of this genus has been controversial and the subject of many reviews. Originally regarded as a subgenus of *Erinaceus*, it was raised to generic status by Corbet (1978, 1988) who also included *Mesechinus* within this genus. *Paraechinus* was considered a distinct genus by Corbet (op. cit.), but a subgenus of *Hemiechinus* by Pavlinov and Rossolimo (1987) and Frost *et al.*, (1991). The latter also elevated *Mesechinus* to a full genus.

SUBGENUS *HEMIECHINUS*

This subgenus consists of two species which are adapted to living in arid climates and which range from the Sahara desert to central Asia. Following Corbet (1978), the Afghanistan form *Hemiechinus auritus megalotis* (given specific status by Ellerman and Morrison-Scott, 1951) is included in *H. auritus*.

Long-eared hedgehog (*H. [Hemiechinus] auritus*)

Taxonomy: *Hemiechinus auritius* Gmelin 1770.

IUCN Category of Threat: Lower Risk (subcategory Least Concern).

Description: The spines are usually banded with dark brown and white and the underparts are generally whitish. In this genus, the ears are longer and more prominent than in other hedgehogs. Body length is approximately 150–270mm and tail length 10–50mm. Unlike *Erinaceus* and *Paraechinus*, *Hemiechinus* lacks a median spineless tract on the top of the head.

The population in Afghanistan and Pakistan is characterised by a very large size (head and body length 260–300mm) and uniformly brown ventral pelage (*H.a. megalotis*) (Corbet, 1992).

Distribution: This species occurs from the eastern Mediterranean (north Libya and Egypt, Cyprus, Asia Minor) through south-west Asia (not Saudi Arabia) to the Gobi desert (Mongolia), Xinjiang (China) and north-west India.

Habitat: *H. auritus* is found in marginal desert regions as well as dry steppes. In Egypt *H. auritus* is rare in poorly vegetated areas but is commonly found in gardens in association with people (Hoogstraal, 1962).

Ecology and behaviour: Nocturnal, terrestrial and solitary animals. During daylight *H. auritus* seeks refuge in a burrow – they are active diggers (Roberts, 1977) – usually under small bushes. In the Punjab of northern India, *H. auritus* may hibernate up to 3.5 months in winter and, in the mountains of Pakistan, hibernation lasts from October to March (Walker, 1991). In warmer areas there is no prolonged winter hibernation but during periods of food scarcity there may be summer aestivation (Roberts, 1977).

Collared hedgehog
(*H. [Hemiechinus] collaris*)

Taxonomy: *Hemiechinus collaris* Gray 1830; *blandfordi* Anderson 1878 is a synonym.

IUCN Category of Threat: Lower Risk (subcategory Least Concern).

Description: Similar to *H. auritius*.

Distribution: Arid zones of north-west India and Pakistan west to the Indus (both banks), north to Jamnur, south as far as Pune (Maharashtra), but the latter may have been an isolate or introduction (Corbet, 1992).

Habitat: Desert and dry steppes.

Ecology and behaviour: Nothing is known about the ecology of this species.

SUBGENUS *PARAECHINUS*

Four species are recognised within this subgenus, all of which are adapted to living in arid regions. Desert hedgehogs feed most often at night, with the diet consisting mainly of insects, but also small vertebrates, eggs of ground-nesting birds and scorpions. Members of this subgenus probably breed only once a year. The range of *Paraechinus* spp. is often sympatric with that of *Hemiechinus* spp.

Desert hedgehog
(*H. [Paraechinus] aethiopicus*)

Taxonomy: *Hemiechinus aethiopicus* Ehrenberg 1833.

IUCN Category of Threat: Lower Risk (subcategory Least Concern).

Description: Similar to *P. micropus*.

Distribution: This species is widely distributed throughout Morocco, Algeria, Tunisia, Egypt, Sudan, Arabia, Aden and Iraq. It is also thought to occur on Gran Canaria, Canary Islands, (Herter, 1972), through introduction (although it may also have been mistaken for *A. algirus*).

Habitat: Desert and dry steppe.

Ecology and behaviour: Little is known about this species. It appears to be marginally sympatric with *Hemiechinus auritus* and *P. hypomelas*.

Brandt's hedgehog
(*H. [Paraechinus] hypomelas*)

Taxonomy: *Erinaceus hypomelas* Brandt 1936. Two subspecies have been recognised in Pakistan: *P.h. hypomelas* (western Pakistan including Baluchistan; Afghanistan) and *P.h. jerdoni*, from the Indus Valley.

IUCN Category of Threat: Lower Risk (subcategory Least Concern).

Description: Similar in external appearance to *P. micropus*. Subspecies may be distinguished on size basis: *P.h. hypomela* is the larger, with a head and body length of 205–285mm, compared to 150–205mm for *P.h. jerdoni*.

Distribution: This species has a wide distribution, ranging from Iran and Turkmenistan to Uzbekistan, to the Indus River and North Pakistan; isolates have been recorded from Oman and on the islands of Tanb and Kharg in the Persian Gulf.

Habitat: Desert, dry steppe.

Ecology and behaviour: More nomadic in habit than other hedgehogs and probably travel over a greater area during the year. Varied diet ranging from insects to snakes and fruit.

Indian hedgehog
(*H. [Paraechinus] micropus*)

Taxonomy: *Hemiechinus micropus* Blyth 1846. A subspecies has been described on the basis of the slight differences in the pigmentation of the spines: *P.m. kutchicus* Biswas and Ghose (1970), who recognised it as belonging to *intermedius*, which they considered a distinct species.

IUCN Category of Threat: Lower Risk (subcategory Least Concern).

Description: Body length varies from 140–230mm and the length of the tail ranges from 10–40mm. Coloration is highly variable; there is a tendency towards melanism and also to albinism. The spines may be banded with dark brown or black and white or yellow, but often just one of these colours predominates. Some forms have a brown muzzle with a white forehead and sides. The underparts may be blotched dark brown and white – the variation ranging from entirely brown to entirely white. The ears are relatively short.

Skull features distinguish this genus on an anatomical basis. *Paraechinus* has a wide and prominent naked area on the scalp, whereas this area is very narrow in *Erinaceus*, moderately wide in *Atelerix*, and lacking altogether in *Hemiechinus* (Corbet, 1988).

Distribution: Arid zones of Pakistan, where it is considered "uncommon", (Roberts, 1977) and northwest India (Hutterer, 1993).

Habitat: Desert and similar arid zones.

Ecology and behaviour: A sedentary, nocturnal species which may seek shelter by day under a pile of brush wood or a bush; it does not always enter burrows (Roberts, 1977) Individuals adhere to their home range for much of their life. *P. micropus* does not hibernate but may remain torpid in its burrow if food or water is scarce (Walker, 1991). The breeding season extends throughout the monsoon period.

Madras hedgehog
(*H. [Paraechinus] nudiventris*)

Taxonomy: *Hemiechinus nudiventris* Horsfield 1851.

IUCN Category of Threat: Lower Risk (subcategory Least Concern).

Distribution: This species has been recorded from Tamil Nadu and Travancore Provinces, India (Corbet, 1992, and Hutterer, 1993, respectively). Additional surveys should be initiated to determine its exact distribution.

Habitat: Desert and similar arid zones.

Ecology and behaviour: The behaviour of this species in the wild is poorly known. Concern has been expressed that it may occupy a restricted distribution.

GENUS *MESECHINUS*

Pavlinov and Rossolimo (1987) placed *Mesechinus* as a subgenus of *Erinaceus*, but Frost *et al.*, (1991) raised it to full generic rank. It contains two species, both found in semi-arid habitats in Mongolia, southern Manchuria, Transbaikalia and North-central China.

Daurian hedgehog
(*Mesechinus dauuricus*)

Taxonomy: *Mesechinus dauuricus* Sundevall 1842.

IUCN Category of Threat: Lower Risk (subcategory Least Concern).

Description: Similar to *H. auritius*.

Distribution: Occurs in a semi-arid zone in North China, ranging from Inner Mongolia to West Manchuria, northeast Mongolia and the Transbaikalia and upper Amur Basin in Russia.

Habitat: Dry steppe.

Ecology and behaviour: Nothing is known about the ecology of this species.

Hugh's hedgehog
(*Mesechinus hughi*)

Taxonomy: *Mesechinus hughi* Thomas 1908.

IUCN Category of Threat: Vulnerable (B1 and 2c).

Description: Similar to *H. auritius*. Characterised by darker spines (due to longer terminal dark tips) and a very pale ventral pelage. This species may also be distinguished by cranial features.

Distribution: Known only from Shaanxi and Shanxi Provinces, Central China (Hutterer, 1993).

Habitat: Dry steppe.

Ecology and behaviour: Nothing is known about the ecology of this species.

Sub-family Hylomyinae

GENUS *ECHINOSOREX*

The single species in this genus, *E. gymnura* is a distinctive animal restricted to the Malay Peninsula, Borneo and Sumatra (Indonesia), including Labuan Island.

Moonrat
(*Echinosorex gymnura*)

Taxonomy: *Viverra gymnura* Raffles 1822. Two subspecies have been described: *E.g. alba* from the eastern and southern regions of Borneo and the Kelabit uplands, as well as Sumatra (Indonesia), Peninsular Malaysia and South Thailand (Corbet, 1992; Payne *et al.*, 1985); *E.g. candida* from the western side of Borneo from P. Labuan south to at least Kuching region (Payne *et al.*, 1985).

IUCN Category of Threat: Lower Risk (subcategory Least Concern).

Description: Members of this genus have an exceedingly narrow body, which may be an adaptation for seeking food in narrow crevices. The rough and harsh pelage consists of a short, thick underfur covered by a dense layer of longer, coarse hair. The colour is variable, and is either usually black – the head and shoulders and the distal part of the tail being whitish (Walker, 1975) – or generally white with a sparse scattering of black hairs (Payne *et al.*, 1985). White forms (not albino) have also been recorded (Lekagul and McNeely, 1977). The face is generally marked with black spots or stripes near the eyes. Payne *et al.*, (1985) also note that on Borneo, at least, moonrats from the west (*E.g. candida*) tend to have a greater proportion of black hairs than those from the east (*E.g. alba*), with an intermediate coloration noted from Brunei.

Moonrats weigh from 1000–1400g (occasionally up to 2kg), with a head and body length of 265–445mm and a tail length of approximately 200mm. Females are larger than males (Davis, 1962; Lekagul and McNeely, 1977). The scantily-haired tail reveals that the scales are arranged in rows around the tail, except near the base where they are arranged diagonally.

Distribution: This monotypic species is found in Tenasserim (Myanmar), Peninsular Thailand, Malaysia, Sumatra and Borneo. Moonrats are widely distributed within their geographical range. They occur in several protected areas, including Tabin Wildlife Reserve (1220km^2) and Danum Valley Conservation Area (430km^2) in Sabah (Payne, pers. comm.).

Habitat: Lowland forests (including logged and secondary), often near streams and mangrove swamps. The habitat of the moonrat ranges into mangrove forest, forest fringe areas and occasionally rubber estates adjacent to secondary forest.

Ecology and behaviour: Nocturnal and terrestrial, remaining in burrows (or rock crevices) during daylight. Earthworms and arthropods are favoured prey in the wild, with fish, crabs and land molluscs serving as supplementary foods (Lim, 1967). Lekagul and McNeely (1977) record that the moonrat often enters water to hunt for frogs, fish, crustaceans, molluscs, and insects. Whitrow, Gould and Rand (1977) suggested that wild specimens eat the fruit of cultivated oil palm.

Little is known about the breeding habits of this species. Lekagul and McNeely (1977) reported that breeding occurs throughout the year with usually two litters per annum, an average litter size of two and a gestation period of 35–40 days. Medway (1978) recorded pregnancies in May, June, September, and November; an average litter size of 1.9 and a record life span in captivity of 55 months.

Although little information is available on this species in the wild, evidence does suggest that moonrats are solitary animals, highly intolerant of conspecifics. In captivity and in the wild, moonrats engage in widespread scent-marking behaviour (see Gould, 1978). *E. gymnura* has a strong, characteristic odour that emanates from two small glands near the anus and that has been variously described as resembling rotten onions, stale sweat and even "Irish stew gone bad"!

GENUS *HYLOMYS*

The genus *Hylomys* (gymnures or moonrats) is widespread throughout south-east Asia, although it is highly fragmented with many isolates. This genus has often been restricted to just a single species, *H. suillus*, although in his extensive revision of the Insectivora, Hutterer (1993) recognised three species. More recently, Ruedi *et al.* (1994) have suggested a fourth species, *H. parvus*, from Sumatra (Indonesia).

Hainan gymnure
(*Hylomys hainanensis*)

Taxonomy: Formerly *Neohylomys hainanensis* Shaw and Wong 1959. The type specimen was collected at Paisa Hsian, Hainan. *Neohylomys* was included within *Hylomys* by van Valen (1967). This arrangement was accepted by Corbet and Hill (1986), Frost *et al.*, (1991) and, with question, by Heaney and Morgan (1982), but not by Corbet (1988; 1992), Honacki, Kinman and Koeppl (1982) or Yates (1984).

IUCN Category of Threat: Endangered (B1 and 2c).

Description: The length of the head and body of seven known specimens range from 120–147mm; the length of the tail from 36–43mm and body weight from 50–69g. The head is blackish-grey, mixed with brown. The back is a rust-grey colour and there is a long, black stripe down the middle of the back. The sides of *H. hainanensis* are washed with olive-yellow and the under parts are pale grey or yellowish-white. The ears, feet and tail are almost naked, with minute scattered short hairs.

The distinguishing features of this species include its generally larger size compared to other *Hylomys* spp., longer tail and the reduction of one lower premolar.

Distribution: *H. hainanensis* is restricted to Hainan Island, off southern China.

Habitat: According to Corbet (1992) this species has been recorded in tropical rainforest and subtropical evergreen forest. It was originally described as subterranean but it is more likely that it merely uses burrows as refuges. Remaining patches of evergreen forest on Hainan Island are now under considerable pressure from clearance for timber and the expansion of agriculture.

Ecology and behaviour: The ecology of this species in the wild is not known.

Dwarf gymnure
(*Hylomys parvus*)

Taxonomy: *H. parvus* Robinson and Kloss 1916. Not regarded as a valid species by Chasen (1940), Corbet (1988), Corbet and Hill (1992), or Hutterer (1993). But see Ruedi *et al.* (1994).

IUCN Category of Threat: Critically Endangered (B1 and 2c).

Description: Like *H. suillus*, but may be separated from the latter by its notched space between premaxillary tips, soft texture of the fur, and more delicate skull and dentition (Ruedi *et al.*, 1994).

Distribution: This species has only been reported from the slopes of Mt Kerinci, West Sumatra, Indonesia.

Habitat: Robinson and Kloss (1918) report finding *H. suillus* and *H. parvus* together at 2200m on Mt Kerinci. Ruedi *et al.* (1994) state that the habitat of the dwarf gymnure is now restricted to the moss forest covering the peak of Mt Kerinci.

Ecology and behaviour: The ecology of this species in the wild is not known. Sympatric with *H. suillus* in the lower part of its range.

Hylomys sinensis

Taxonomy: Formerly *Neotetrachus sinensis* Trouessart 1909. *N. sinensis* was synonymised with *Hylomys* by van Valen (1967). This arrangement was accepted by Corbet and Hill (1986) and Frost *et al.*, (1991), but not by Corbet (1988, 1992), Heaney and Morgan (1982), Honacki, Kinman and Koeppl (1982) or Yates (1984). According to Corbet (1992), five nominal subspecies have been described: *N.s. sinensis*, *N.s. flavescens*, *N.s. cuttingi*, *N.s. hypolineatus* and *N.s. unicolor*.

IUCN Category of Threat: Lower Risk (subcategory Near Threatened).

Description: The length of the head and body is 105–148mm; tail length varies from 60–82mm. The coat is soft, dense and quite long. In appearance, the dorsal surface may be olive-brown, cinnamon-brown, or a mixed cream colour and black, with the sides of the head and neck sometimes tinged with red. An indistinct, blackish dorsal stripe may be present. The underparts are reddish, buff-grey or cream-coloured over a dark background.

H. sinensis is distinguished from other members of this genus by its longer tail, shorter snout and fewer teeth. The tail of *H. sinensis* is also thinly covered with minute hairs. Females have eight mammae.

Distribution: Montane areas of Sichuan and Yunnan (China) and adjacent parts of Myanmar and North Vietnam.

Habitat: It inhabits cool damp forests between 300 and 2700m in altitude.

Ecology and behaviour: These animals are apparently strictly terrestrial and nocturnal. They appear quite common in parts of their range and are found in runways and burrows with moss and fern cover, as well as beneath logs and rocks. Diet appears to be composed mostly of invertebrates. The breeding season appears to extend throughout the year, probably limited to two litters per annum.

Lesser gymnure
(*Hylomys suillus*)

Taxonomy: *Hylomys suillus* Müller 1840. Three distinct subspecies can be recognised (Corbet, 1992): *H.s. suillus* from Java and Sumatra, Indonesia (no dorsal stripe); *H.s. dorsalis* from Borneo (slight mid-dorsal stripe) and *H.s. tionis* from Tioman Island, Malaysia (no dorsal stripe).

IUCN Category of Threat: Lower Risk (subcategory Least Concern).

Description: The length of the head and body is approximately 105–146mm, while tail length ranges from 12–30mm. The pelage is soft and rusty brown in colour on the dorsal surface, grey or yellowish on the ventral side. An indistinct black nape strip or black dorsal stripe may be present (see Taxonomy).

Distribution: The lesser gymnure is found along the Myanmar border of Yunnan, China, in Myanmar, Indochina, Thailand, Peninsular Malaysia, and on Tioman Island, Sumatra, Java and Borneo. It is known to occur in Kinabalu National Park (750km^2), Sabah.

Habitat: *H. suillus* is confined to forest areas. Its preferred habitat appears to be humid mountain or lowland forest with thick undergrowth. It has been recorded from an altitude of about 90m on the mainland, 3000m on Sumatra and 1000–3400m on Borneo.

Ecology and behaviour: Although capable of climbing, this species is generally terrestrial, moving in short bounds and bursts of speed when threatened. It appears to use regular paths on the forest floor. Diet is mainly composed of invertebrates such as insects and earthworms. It may also feed on fruit. *H. suillus* is active at infrequent periods during the day and night. Breeding appears to take place throughout the year, with 2–3 young born after a gestation of 30–35 days. Longevity probably does not exceed two years. As in *Echinosorex*, a strong odour is characteristic of this species (Lekagul and McNeely, 1977; Walker, 1991).

Lesser gymnures probably breed throughout the year, at least in the tropical parts of their range. Shelter is sought in nests of dead leaves made in hollows in the ground or under rocks.

GENUS *PODOGYMNURA*

The genus *Podogymnura* contains two species, both of which are restricted to the southern Philippines. Their taxonomy has been reviewed by Heaney and Morgan (1982) and Poduschka and Poduschka (1985).

Dinagat moonrat
(*Podogymnura aureospinula*)

Taxonomy: *Podogymnura aureospinula* Heaney and Morgan 1982; described from Dinagat Island, the Philippines.

IUCN Category of Threat: Endangered (B1 and 2c).

Description: Distinguished by its stiff dorsal pelage, which is generally golden brown in colour, with black speckling. Underparts lack stiff hairs and are mostly brownish-grey.

Distribution: Restricted to Dinagat Island, the Philippines.

Habitat: The few known specimens were collected from logged dipterocarp forest.

Ecology and behaviour: Nothing is known about the ecology of this species.

Mindanao moonrat
(*Podogymnura truei*)

Taxonomy: *P. truei* Mearns 1905; described from Mt Apo, Mindanao.

IUCN Category of Threat: Endangered (B1 and 2c).

Description: The length of head and body is approximately 130–150mm, while tail length varies from 40–70mm. The pelage is long, soft and full. Dorsal coloration is predominantly grey mixed with reddish-brown hairs, while the under parts are hoary, slightly mixed with brown hairs. The tail is partially furred and is a buff-purple colour.

Distribution: This species is only known from Mindanao in the Philippine Islands. Also known as the Philippines gymnure and Bagobo ("ground-pig") by natives, it has been collected on Mount Apo at elevations of 1700–2100m, on the eastern slope of Mt McKinley from 1800–2300m and on Mt Katanglad at an elevation of 1600m.

Habitat: Probably confined to forest, frequenting areas of standing water.

Ecology and behaviour: Almost nothing is known about the life history of this species. Habits are comparable to those of true shrews (Walker, 1991).

2.3.2 Family Soricidae: The Shrews

The Family Soricidae consists of 23 genera and about 314 species, including 12 genera and 126 species in Eurasia (Hutterer, 1993) (Table 2.2). This is an ancient family, diverging from other insectivores before the Eocene period, with modern genera first appearing in the Miocene. Originating in Eurasia, the soricids later migrated to Africa and North America. One genus (*Cryptotis*) has even reached South America.

Table 2.2. General classification of Eurasian Soricidae (Hutterer, 1993)
(see Tables 2.3 and 2.4 for further details)

Sub-family	Genus
Crocidurinae	*Crocidura*
	Diplomesodon
	Feroculus
	Solisorex
	Suncus
Soricinae	*Anourosorex*
	Blarinella
	Chimarrogale
	Nectogale
	Neomys
	Sorex
	Soriculus

Shrews are the smallest of the insectivores: one species, *Suncus etruscus* of southern Europe and Asia, is one of the smallest known mammals. All have short legs, five clawed toes, a relatively long tail (in most genera), short dense fur, small external ears and an elongated snout. The eyes are small. All are insectivorous or carnivorous, living on the ground in leaf litter and grass. Shrews are voracious animals with little resistance to starvation. As a result, most are active for short periods of time throughout the day and night. Some species may even undergo torpor when local environmental conditions (notably climate and food shortages) are unfavourable.

The Soricidae are divided into two sub-families: the Crocidurinae (white-toothed shrews) and the Soricinae (red-toothed shrews) (see Tables 2.3 and 2.4 for further taxonomic details). Red-toothed shrews are so called because their teeth have a reddish appearance on account of a deposition of iron in the outer layer of enamel (Dötsch and Koenigswald, 1978), which may increase resistance to wear.

Despite considerable attention, the taxonomic status of the Soricidae is still unclear, particularly for some genera. Modern techniques, however, have helped clarify a number of dubious relationships. Yet, as the following accounts demonstrate, there is still a need for considerable research, both at the taxonomic and ecological levels. For many species no details of their ecological requirements are available, many being known only from a single location and few records.

Sub-family Crocidurinae

GENUS *CROCIDURA*

The genus *Crocidura* ('white-toothed shrews') is a widespread and highly variable Old World genus, which is found in many African countries, much of continental Europe and parts of south-eastern Asia. This genus comprises 151 species (55 of which are represented in Eurasia), which are distinguished from other shrews (sub-family Soricinae) by their unpigmented teeth, the presence of three upper unicuspids, long scattered hairs on the tail, and more prominent ears than in either the *Sorex* or *Neomys* genera.

Crocidura aleksandrisi

Taxonomy: *Crocidura aleksandrisi* Vesmanis 1977.

IUCN Category of Threat: Lower Risk (subcategory Least Concern).

Distribution: This species is restricted to Cyrenaica, Libya.

Ecology and behaviour: Although little is known about the ecology of this species, there is no indication that it is threatened (Hutterer, pers. comm.).

Crocidura andamanensis

Taxonomy: *Crocidura andamanensis* Miller 1902. This species has been described on the basis of a single individual.

IUCN Category of Threat: Endangered (B1 and 2c).

Distribution: This species has only been recorded from South Andaman Island, Indian Ocean.

Crocidura arabica

Taxonomy: *Crocidura arabica* Hutterer and Harrison 1988. Previously assigned to *C. russula* or *C. suaveolens* (Hutterer, 1993).

IUCN Category of Threat: Lower Risk (subcategory Least Concern).

Table 2.3. Classification of the Eurasian Soricidae – Sub-family Crocidurinae (Hutterer, 1993)			
Genus	**Species**	**Genus**	**Species**
Crocidura	C. aleksandrisi	Crocidura (cont.)	C. osorio
	C. andamanensis		C. palawanensis
	C. arabica		C. paradoxura
	C. armenica		C. pergrisea
	C. attenuata		C. pullata
	C. beatus		C. religiosa
	C. beccarii		C. rhoditis
	C. canariensis		C. russula
	C. dhofarensis		C. sereckyensis
	C. dsinezumi		C. shantungensis
	C. elongata		C. sibirica
	C. floweri		C. sicula
	C. fuliginosa		C. suaveolens
	C. gmelini		C. susiana
	C. grandis		C. tenuis
	C. grayi		C. whitakeri
	C. gueldenstaedtii		C. zarudni
	C. hispida		C. zimmermanni
	C. horsfieldii		
	C. jenkinsii	Diplomesodon	D. pulchellum
	C. lasiura		
	C. lea	Feroculus	F. feroculus
	C. leucodon		
	C. levicula	Solisorex	S. pearsoni
	C. malayana		
	C. maxi	Suncus	S. ater
	C. mindorus		S. dayi
	C. minuta		S. etruscus
	C. miya		S. fellowsgordoni
	C. monticola		S. hosei
	C. neglecta		S. malayanus
	C. negrina		S. mertensi
	C. nicobarica		S. montanus
	C. nigriceps		S. murinus
	C. olivieri		S. stoliczkanus
	C. orientalis		S. zeylanicus
	C. orii		

Distribution: This species is known only from the coastal plains of the South Arabian Peninsula, specifically Yemen and Oman.

Crocidura armenica

Taxonomy: *Crocidura armenica* Gureev 1963. Considered conspecific with *C. pergrisea* by Corbet (1978), but see Dolgov and Yudin (1975) and Gromov and Baranova (1981).

IUCN Category of Threat: Data Deficient.

Distribution: *C. armenica* has been recorded from Armenia, Caucasus.

Grey shrew
(*Crocidura attenuata*)

Taxonomy: *Crocidura attenuata* Milne-Edwards 1872. Inclusion of *C. aequicaudata* as a subspecies may not be justified (Hutterer, 1993).

IUCN Category of Threat: Lower Risk (subcategory Least Concern).

Description: In appearance, the grey shrew is a uniform light brownish-grey above, a paler grey below and faintly tinged with brown on the underside. The tail is dark brown above, light below. The backs of the feet are thinly covered with short pale hairs (Lekagul and McNeely, 1977).

Distribution: This species has been recorded in southern China from Sichuan, Hubei and Jiangsu south to Myanmar and Malaysia (Davison, 1984) and the Himalayas west to Kashmir and Pakistan; Thailand (McNeely, pers. comm.); Taiwan and Hainan, as well as Batan Island, the Philippines. It may also occur on Sumatra (Jenkins, 1982) and was perhaps formerly present on Christmas Island, Indian Ocean (*trichura*).

Habitat: Thai specimens have been recorded from seedling rice fields and in cut-down forest-farmlands of weeds and grass (Lekagul and McNeely, 1977).

Crocidura beatus

Taxonomy: *Crocidura beatus* Miller 1910. Includes *parvacauda* (Heaney *et al.*, 1987).

IUCN Category of Threat: Vulnerable (B1 and 2c).

Distribution: *Crocidura beatus* has been recorded from the islands of Mindanao, Leyte and Maripipi, the Philippines.

Habitat: This species is common in primary forest, uncommon in secondary forest and absent outside of forest (Heaney *et al.*, 1987). It may be endangered by habitat change.

Crocidura beccarii

Taxonomy: *Crocidura beccarii* Dobson 1886.

IUCN Category of Threat: Endangered (B1 and 2c).

Distribution: This species has only been recorded from Mt Singgalang, West Sumatra.

Crocidura canariensis

Taxonomy: *Crocidura canariensis* Hutterer, Lopez-Jurado and Vogel 1987.

IUCN Category of Threat: Vulnerable (B1 and 2c). Protected by Spanish law.

Description: This species may be separated from the other endemic shrew *C. osorio* by its larger size and the uniformly chocolate brown undersurface of the former.

Distribution: This species has been described from the east Canary islands – Fuerteventura, Lanzarote and Lobos.

Habitat: The main habitat for this species is the *malpaís* (barren lava fields); *C. canariensis* appears to be adapted to the hot and dry conditions of these plains (Hutterer *et al.*, 1992).

Ecology and behaviour: Little is known about the behaviour of this species in the wild. Hutterer *et al.*, (1992) describe it as being a "shy animal which will always hide under shelter." Limited reproductive data are given in Hutterer *et al.*, (loc. cit.). It is of interest to note that this species, together with the endemic *C. osorio* have very low litter sizes, compared to mainland species. According to Genoud (1988) a small litter size may be part of a strategy for living under warm and/or unpredictable conditions, which includes a lower rate of body metabolism. Volcanic islands, such as the Canary Islands, represent such unpredictable environments.

This species, together with *C. osorio*, represents two-thirds of the extant native Canarian mammalian fauna – the other species being a bat, *Plecotus teneriffae* (Ibañez and Fernández, 1985). As such, they are a major part of the country's natural heritage and therefore worthy of special protection. Greater emphasis should therefore be given to determining the ecological needs and conservation requirements of these species. Both *C. osorio* and *C. canariensis* are now protected by Spanish law (Hutterer, 1993).

Crocidura dhofarensis

Taxonomy: *Crocidura dhofarensis* Hutterer and Harrison 1988. Considered a subspecies of *C. somalica*.

IUCN Category of Threat: Critically Endangered (B1 and 2c).

Distribution: Known only from the type locality, Oman, Dhofar, Khadrafi (Hutterer and Harrison, 1988).

Japanese white-toothed shrew (*Crocidura dsinezumi*)

Taxonomy: *Crocidura dsinezumi* Temminck 1844.

IUCN Category of Threat: Lower Risk (subcategory Least Concern).

Description: Head and body length is usually 65–80mm; tail length is usually less than 70% of head and body length. Colour of pelage is variable; during the winter, the dorsal colour ranges from pale grey to brown, while summer coloration is usually dark brown. These shrews are usually a lighter colour on the underside.

Distribution: This species is found in Japan, especially Honshu and the islands of Kyushu, Shikoku, Yakushima,

Tanegashima, Oki and Okinoshima. It has also been reported from Quelpart Island, South Korea and may occur on Taiwan (Hutterer, 1993).

Habitat: This species is most abundant along river banks and in the foothills with dense vegetation (Abe, pers. comm.).

Ecology and behaviour: Apart from its diet – insects and spiders – little is known about the behaviour of this species in the wild.

Crocidura elongata

Taxonomy: *Crocidura elongata* Miller and Hollister 1921.

IUCN Category of Threat: Lower Risk (subcategory Least Concern).

Distribution: This species has only been recorded from North and Central Sulawesi, Indonesia.

Habitat: Lowland and montane forest.

Ecology and behaviour: This species is thought to be primarily nocturnal. No other details are available on its behaviour or ecology.

Flower's shrew (*Crocidura floweri*)

Taxonomy: *Crocidura floweri* Dollman 1915.

IUCN Category of Threat: Endangered (B1 and 2c).

Distribution: This species is only known from El Giza and the Nile Delta, Egypt.

Crocidura fuliginosa

Taxonomy: *Crocidura fuliginosa* Blyth 1855. At least 19 subspecies are mentioned in the literature: including *C.f. baluensis* which may be a separate species (Payne *et al.*, 1985); but see Hutterer (1993).

IUCN Category of Threat: Lower Risk (subcategory Least Concern).

Description: *C. fuliginosa* has dark grey to blackish fur with a dull silvery gloss. Underparts are a lighter colour. The tail is thin with a few faint white hairs. The basal portion of the tail is thickest during the breeding season (Blanford, 1888). Ears are naked and prominent. Eyes are small. The feet are covered with a few short white hairs (Lekagul and McNeely, 1977).

Distribution: This widely dispersed species has been recorded from China to Malaya; it may also occur on Sumatra and Java (Indonesia), Borneo; adjacent small islands of Hong Kong, Con Son (Vietnam); Samui (Thailand); Great Redang, Tioman, Aor, Dayang Bunting (Malaya); Mapor (Riau Island) and Panjang (South Natuna Island). Ruedi (1994), however, suggests that *C. fuliginosa* is purely a continental species which enters only marginally into the south-east Asian archipelago.

Habitat: Found in various habitats from montane to lowland forest, cultivated areas and even caves (Lekagul and McNeely, 1977).

Crocidura gmelini

Taxonomy: *Crocidura gmelini* Pallas 1811. The name has traditionally been considered a synonym of *Sorex minutus*, but Hoffmann (in press) has shown that it is applicable to a distinct species of *Crocidura* that inhabits steppe and semi-desert from Central Iran to Central China, and which had previously been considered conspecific with *C. suaveolens*.

IUCN Category of Threat: Lower Risk (subcategory Least Concern).

Distribution: Central Iran to Central China.

Habitat: Steppe and semi-desert conditions.

Crocidura grandis

Taxonomy: *Crocidura grandis* Miller 1911.

IUCN Category of Threat: Endangered (B1 and 2c).

Distribution: The only records for this species are from Mt Malindang, Mindanao, the Philippines.

Habitat: Probably confined to primary forest (Heaney *et al.*, 1987).

Ecology and behaviour: No information available.

Crocidura grayi

Taxonomy: *Crocidura grayi* Dobson 1890. Includes *C. halconus* Heaney *et al.*, 1987.

IUCN Category of Threat: Vulnerable (B1 and 2c).

The rare *Crocidura grayi* is only known from the islands of Luzon and Mindoro, the Philippines. (Photo by Andres L. Dans)

Distribution: This species occurs only on Luzon and Mindoro in the Philippines.

Crocidura gueldenstaedtii

Taxonomy: *Crocidura gueldenstaedti* Pallas 1811. This form was previously included in the European *C. russula* by Ellerman and Morrison-Scott (1951) but is now generally agreed to be a distinct species. However, it is likely that some or all of these forms are conspecific with *C. suaveolens* Pallas (Catzeflis *et al.*, 1985; Hutterer, 1993).

IUCN Category of Threat: Lower Risk (subcategory Least Concern).

Distribution: The distribution of this species is somewhat unclear because of remaining taxonomic uncertainties. Its apparent range extends from eastern Europe through Central and south-western Asia to China and Taiwan.

Andaman shrew
(*Crocidura hispida*)

Taxonomy: *Crocidura hispida* Thomas 1913.

IUCN Category of Threat: Endangered (B1 and 2c).

Distribution: This species has only been recorded on the Middle Andaman Island, Indian Ocean.

Habitat: The preferred habitat of this species is unknown.

Horsfield's shrew
(*Crocidura horsfieldii*)

Taxonomy: *Crocidura horsfieldii* Tomes 1856. Two insular subspecies have been described: *C.h. wuchihensis* from the western slope of Wuchih Mountain, Hainan Island, China and *C.h. kurodai* (Jameson and Jones, 1977) from Linkou, Taipai Hsein, Taiwan.

IUCN Category of Threat: Lower Risk (subcategory Least Concern).

Description: *C. horsfieldii* is a deep neutral grey above with the bottom portion of the hairs brown. Underparts are dark grey. The tail is paler above than beneath, with short scattered silver hairs (Robinson and Kloss, 1922).

Distribution: The exact distribution of this species is poorly known. It is known to occur in Sri Lanka, South India, Kashmir, North Myanmar, Thailand, Vietnam, Hainan, Botel Tobago (south-eastern Taiwan) and the Ryukyu Islands of Okinawa, Yoron, Iejima, Okinoerabu, Tokunoshima and Amani-Oshima (Corbet, 1992).

Habitat: This species is an intermediate montane form which has been collected from 1220–2120m in areas of fairly heavy cover (Lekagul and McNeely, 1977).

Crocidura jenkinsii

Taxonomy: *Crocidura jenkinsi* Chakraborty 1978. May be a subspecies of *C. nicobarica*.

IUCN Category of Threat: Critically Endangered (B1 and 2c).

Distribution: This species is only known from the type locality at Wright Myo, South Andaman Island, India.

Ussuri white-toothed shrew
(*Crocidura lasiura*)

Taxonomy: *Crocidura lasiura* Dobson 1890.

IUCN Category of Threat: Lower Risk (subcategory Least Concern).

Distribution: Confined to the Ussuri region of the former Soviet Union and Manchuria south to north-eastern China and Korea.

Crocidura lea

Taxonomy: *Crocidura lea* Miller and Hollister 1921.

IUCN Category of Threat: Lower Risk (subcategory Least Concern).

Distribution: North and Central Sulawesi, Indonesia.

Habitat: Lowland and montane forest.

Ecology and behaviour: This species is believed to be primarily nocturnal in habit. There are no additional data available at the present time.

Bicoloured white-toothed shrew (*Crocidura leucodon*)

With a wide distribution in central and eastern Europe the bicoloured white-toothed shrew (*Crocidura leucodon*) has a patchy distribution in the west. (Photo by Peter Vogel)

Taxonomy: *Sorex leucodon* Hermann.

IUCN Category of Threat: Lower Risk (subcategory Least Concern).

Description: *C. leucodon* may be distinguished from *C. russula* and *C. gueldenstaedtii* by having a sharper line of demarcation between the dorsal and ventral pelage.

Distribution: This species is found in much of central and eastern Europe with the exception of the British Isles, Iberia, southern and western France, northern Europe, Italy and Greece. An isolated population occurs in southern Italy. Towards the east, it ranges as far as Asia Minor, Arabia and Iran and has been recorded in the Caucasus and Elburz mountains. It is unlikely to occur further east (Corbet, 1992). This species is listed as 'Endangered' in the states of Mecklenburg-Vorpommern and Schleswig-Holstein, Germany (Ingelög *et al.*, 1993).

Habitat: As for *C. russula*.

Ecology and behaviour: This species has been poorly studied in the wild. It exhibits caravanning behaviour, whereby an adult leads its offspring, each holding in its mouth the tail or rump of the animal in front. This may occur when a nest is disturbed or when juveniles first begin to forage with their mother.

Crocidura levicula

Taxonomy: *Crocidura levicula* Miller and Hollister 1921.

IUCN Category of Threat: Lower Risk (subcategory Least Concern).

Distribution: Central and south-eastern Sulawesi.

Habitat: Lowland and montane forest.

Ecology and behaviour: *Crocidura levicula* is a diurnal species. No other information is available concerning its ecology or behaviour.

Crocidura malayana

Taxonomy: *Crocidura malayana* Robinson and Kloss 1911.

IUCN Category of Threat: Endangered (B1 and 2c).

Distribution: The exact distribution of this species is unknown; it has been recorded from Perak (west Peninsular Malaysia) and offshore islands.

Crocidura maxi

Taxonomy: *Crocidura maxi* Sody 1936. Ruedi (1994) has suggested that this species be reclassified as *C. monticola*.

IUCN Category of Threat: Lower Risk (subcategory Least Concern).

Distribution: Recorded from Java, the Lesser Sunda Islands and Ambon (Indonesia).

Crocidura mindorus

Taxonomy: *Crocidura mindorus* Miller 1910.

IUCN Category of Threat: Endangered (B1 and 2c).

Distribution: This species has been recorded from Mt Halcon, Mindoro, Sibuyan Island and the north-west slope of Mt Guiting-guiting, the Philippines (Heaney and Ruedi, 1994).

Crocidura minuta

Taxonomy: *Crocidura minuta* Otten 1917. The status of this name is uncertain; it may be a subspecies of *C. monticola*, or a senior synonym for *C. maxi*; see Hutterer (1993).

IUCN Category of Threat: Data Deficient.

Distribution: This species has only been reported from East Java, Indonesia.

Sri Lankan long-tailed shrew (*Crocidura miya*)

Taxonomy: *Crocidura miya* Phillips 1929.

IUCN Category of Threat: Endangered (B1 and 2c).

Distribution: This species has been described from the central highlands of Sri Lanka, at an altitude of approximately 1000–2000m.

Crocidura monticola

Taxonomy: *Crocidura monticola* Peters 1870.

IUCN Category of Threat: Lower Risk (subcategory Least Concern).

Description: Body uniform dull grey-brown. Tail paler with sparse, long, pale hair at the base, usually extending over 1cm along the tail.

Distribution: This species has been recorded from Java (Indonesia), Borneo and Peninsular Malaysia (Hutterer, 1993). It has also been observed on Ambon Island (Indonesia) and may exist on the following Indonesian islands: Sumbawa, Lombok, Sumba, Flores, Komodo, Obi and Timor, but all of these need confirmation (Jenkins, 1982).

Crocidura neglecta

Taxonomy: *Crocidura neglecta* Jentink 1888. May be a senior synonym for *C. maxi*. Ruedi (1994) has suggested that this species be reclassified as *C. monticola*.

IUCN Category of Threat: Lower Risk (subcategory Least Concern).

Distribution: This species has been recorded from Indonesia, specifically from Sumatra (perhaps including Mt Kerinci), East Java, Sumba, Flores and perhaps Ambon (Moluccas) (Jenkins, 1982). It is sympatric with *C. monticola* on Java.

Crocidura negrina

Taxonomy: *Crocidura negrina* Rabor 1952.

IUCN Category of Threat: Critically Endangered (B1 and 2c).

Description: Small, closely resembling *Suncus murinus* in external appearance, but much smaller with a relatively longer and more slender tail and much darker pelage; general colour is blackish, tinged with grey on the upperparts; dark hair, brown on underparts with numerous whitish spots, especially on the belly (Rabor, 1977).

Distribution: This species has only been recorded from Cuernos de Negros mountain, Negros Island, the Philippines. Six specimens are known from primary forest at an altitude range of 500–1450m.

Nicobar shrew (*Crocidura nicobarica*)

Taxonomy: *Crocidura nicobarica* Miller 1902. May include *C. jenkinsi* (Hutterer, 1993).

IUCN Category of Threat: Endangered (B1 and 2c).

Distribution: This species has only been recorded from Great Nicobar Island in the Bay of Bengal.

Crocidura nigripes

Taxonomy: *Crocidura nigripes* Miller and Hollister 1921. A single subspecies, *C.n. lipari*, has been described on the basis of its larger size.

IUCN Category of Threat: Lower Risk (subcategory Least Concern).

Distribution: North and Central Sulawesi, including the island of Lembeh (north-east Sulawesi).

Habitat: Lowland forest.

Ecology and behaviour: Field observations suggest that this species is primarily nocturnal. No other details of its ecology or behaviour are available at the present time.

Crocidura olivieri

Taxonomy: *Crocidura olivieri* Lesson 1827.

IUCN Category of Threat: Lower Risk (subcategory Least Concern).

Distribution: This species has been recorded from Egypt. Its range also extends from Senegal to Ethiopia and south to northern South Africa.

The African giant shrew (*Crocidura olivieri*) from Bangui, Central African Republic. This species has been recorded from Egypt. (Photo by Tiziano Maddalena)

Crocidura orientalis

Taxonomy: *Crocidura orientalis* Jentink 1890. Previously included with *C. fuliginosa*, but described as a separate species by Corbet (1992) and Ruedi (1994).

IUCN Category of Threat: Vulnerable (D2).

Description: This species is distinguished from the probably parapatric forms of *C. fuliginosa* by its longer tail, absence of caudal vibrissae, narrower, much less robust teeth and more pointed coronoid and slender angular processes of the mandibles. It is also smaller, on average, but there is a slight degree of overlap in size (Corbet, 1992).

Distribution: Known for certain only from the vicinity of the type locality at Cibodas, western Java (Indonesia), including Mt Pangrango and Mt Gede at altitudes of 1800–2700m (Corbet, 1992).

Crocidura orii

Taxonomy: *Crocidura orii* Kuroda 1924. Considered a subspecies of *C. dsinesumi* by Corbet (1978); but see Hutterer (1993).

IUCN Category of Threat: Endangered (B1 and 2c).

Description: Closely resembles *C. dsinezumi*, but *C. orii* is generally larger.

Distribution: This species is known only from the islands of Amani-Oshima and Tokunoshima in the Ryukyu Islands where it is sympatric with the smaller *C. horsfieldii*.

Habitat: The habitat preferences of this species are unknown at present.

Crocidura osorio

Taxonomy: *Crocidura osorio* Molina and Hutterer 1989.

IUCN Category of Threat: Vulnerable (D2). This species is protected by Spanish law (Hutterer, 1993).

Description: A small shrew with an average head and body length of 55mm. The general colour is grey-brown, but the species may be identified from the other endemic shrew *C. canariensis*, by the white undersurface of the former species. There is no sharp distinction between dorsal and ventral coloration (Molina and Hutterer, 1989).

Distribution: This species has only been recorded from the northern cloud zone of Gran Canaria, Canary Islands, Spain. The known distribution of this species is extremely small. Rapid urbanisation, together with changes caused by increasing desiccation on the island may threaten the species (Molina and Hutterer, 1989).

Habitat: Humid parts of northern Gran Canaria where it occurs in remnant patches of the former laurisilva belt, as well as extensively managed farmland (Molina and Hutterer, 1989).

Ecology and behaviour: A recently described species, little is known about this shrew in the wild. One of three endemic mammals to the island, the ecology of this species should be investigated in greater detail (see also comments for *C. canariensis*, which apply to *C. osorio* as well).

Crocidura palawanensis

Taxonomy: *Crocidura palawanensis* Taylor 1934. May be conspecific with *C. fuliginosa* (Heaney *et al.*, 1987).

A threatened species, the Palawan tree shrew (*Tupaia palawanensis*) occurs only in the Philippines. (Photo by Andres L. Dans)

IUCN Category of Threat: Vulnerable (B1 and 2c).

Distribution: This species has only been recorded from Palawan, the Philippines.

Crocidura paradoxura

Taxonomy: *Crocidura paradoxura* Dobson 1886.

IUCN Category of Threat: Endangered (B1 and 2c).

Description: This species has been described from the holotype which is characterised by a relatively longer tail (159% of head and body length) than in any other Indomalayan *Crocidura*.

Distribution: This species has only been observed on Mt Singgalang, West Sumatra, at an altitude of 2000m. It may also occur on Mt Pangrango in West Java (1450m) (Corbet, 1992).

Pale grey shrew (*Crocidura pergrisea*)

Taxonomy: *Crocidura pergrisea* Miller 1913. Formerly included *armenica*, *serezkyensis* and *zarudnyi*, which are now considered distinct species (Hutterer, 1993).

IUCN Category of Threat: Vulnerable (B1 and 2c).

Description: A medium-sized shrew of pale grey coloration. Average size: body and head length 56mm; tail length 45mm.

Distribution: This species is only known from Kashmir, specifically the sites of Baltistan, Shigar and Skoro Loomba.

Habitat: Montane; adapted to semi-arid conditions.

Crocidura pullata

Taxonomy: *Crocidura pullata* Miller 1911. Formerly included in *C. russula* by Jenkins (1976); but see Hutterer (1993), who provisionally included *rapax* and *vorax* as synonyms.

IUCN Category of Threat: Lower Risk (subcategory Least Concern).

Distribution: The extent of this species' range is unknown, but includes Kashmir, India, Afghanistan, Pakistan, Yunnan (China) and Thailand.

Egyptian pygmy shrew (*Crocidura religosa*)

Taxonomy: *Crocidura religosa* I. Geoffroy 1827. Considered conspecific with *C. nana*, but see Hutterer (1993).

IUCN Category of Threat: Data Deficient.

Distribution: This species has only been recorded from the Nile Valley, Egypt.

Habitat: The preferred habitat of this species is not known.

Crocidura rhoditis

Taxonomy: *Crocidura rhoditis* Miller and Hollister 1921.

IUCN Category of Threat: Lower Risk (subcategory Least Concern).

Description: Similar to *C. grayi* from the Philippines.

Distribution: This species occurs in North, Central and south-western Sulawesi.

Habitat: Lowland and montane forest.

Ecology and behaviour: Field observations suggest that this species is nocturnal in habit. At present, however, no other details of its ecology are known.

Greater white-toothed shrew (*Crocidura russula*)

Taxonomy: *Sorex russula* Hermann. Populations from eastern Europe to Iran formerly included in *C. russula* (Corbet, 1978) are now assigned to *C. suaveolens* (Catzeflis *et al.*, 1985). One subspecies *C.r. cossyrensis*, described from Pantelleria Island, Italy, has been variously referred to as a full species, *C. cossyrensis* (Contoli, 1990), as a subspecies of *russula* (Contoli and Amori, 1986), and as closely related to *russula* (Sará *et al.*, 1990). More recently, Vogel *et al.*, (1992), suggest that this is in fact a subspecies of African origin. In view of its precarious position, this subspecies has been assigned a separate IUCN Category: Vulnerable (D2).

IUCN Category of Threat: Lower Risk (subcategory Least Concern).

Description: Upper surface greyish- or reddish-brown; underside a yellowish-grey with no clear line of separation.

One of the commonest shrews in Eurasia, the greater white-toothed shrew (*Crocidura russula*) is well adapted to foraging in short grass and woodlands. (Photo by Peter Vogel)

Little seasonal variation in fur colour. Ears prominent and tail furnished with long, white hairs as in lesser white-toothed shrew. Separable from *C. suaveolens* only by dental characters.

Distribution: This species is one of the commonest of the white-toothed shrews in western Europe. Its range extends from Iberia and the Mediterranean, north-east to northern Germany, southern Poland, and in North Africa from Algeria to Morocco.

Habitat: This species inhabits grassland, woodland, and hedgerows, especially on dry ground. Usually below 1000m, although it has been recorded at an altitude of 1600m in the Alps. Often commensal with man, living around houses and out-buildings.

Ecology and behaviour: Active day and night with peaks of activity after dusk and around dawn. Breeding may extend from February to November in the south.

Crocidura serezkyensis

Taxonomy: *Crocidura serezkyensis* Laptev 1929. Formerly included in *C. pergrisea* (Jenkins, 1976), but see Hutterer (1993).

IUCN Category of Threat: Lower Risk (subcategory Least Concern).

Distribution: Asia Minor, Azerbaijan, Turkmenistan, Tadshikistan and Kazakhstan.

Crocidura shantungensis

Taxonomy: *Crocidura shantungensis* Miller 1901. Considered a subspecies of *C. suaveolens* by Corbet (1978), but see Hoffmann (in press).

IUCN Category of Threat: Data Deficient.

Distribution: This species is known only from Shandong Province, China.

Crocidura sibirica

Taxonomy: *Crocidura sibirica* Dukelski 1930.

IUCN Category of Threat: Lower Risk (subcategory Least Concern).

Distribution: Central Asia from Lake Issyk Kul to the Upper Ob River and Lake Baikal; perhaps also Mongolia and Sinkiang, China, (Hutterer, 1993).

Sicilian shrew (*Crocidura sicula*)

Taxonomy: *Crocidura sicula* Miller 1900. Formerly included in *leucodon*, *russula* or *suaveolens* but the species has a distinct karyotype and morphology (Vogel *et al.*, 1989). Four subspecies have been recognised – *estuae*, *sicula*, *aegatensis* and *calypso*, of which *estuae* (known from the Pleistocene from Sicily and Malta) is now extinct (see Hutterer, 1991).

IUCN Category of Threat: Lower Risk (subcategory Least Concern).

Description: This species is characterised by a sharply bicoloured body and tail (see Vogel *et al.*, 1989).

Distribution: This species has only been recorded from Sicily, Malta and Egadi Island in the Mediterranean.

Habitat: Scrub and woodlands with ample ground cover.

This Sicilian shrew (*Crocidura sicula*) from Gozo occurs only on Sicily, Malta and Egadi Island in the Mediterranean. (Photo by Peter Vogel)

The caravanning behaviour of the Sicilian shrew (*Crocidura sicula*) is used when a mother and offspring change their nest site. (Photo by Peter Vogel)

Lesser white-toothed shrew (*Crocidura suaveolens*)

Taxonomy: *Crocidura suaveolens* Pallas 1811.

IUCN Category of Threat: Lower Risk (subcategory Least Concern).

Description: The pelage is grey-reddish brown above and slightly lighter on the underside, with little seasonal variation. Ears are short-haired and prominent. The tail is covered with short bristly hairs interspersed with fine, long, white hairs. Similar to *C. russula* and only distinguishable by careful measurement. Externally, the tail (which measures 24–44mm in *C. suaveolens*; 33–46mm in *russula*) and hind feet (10–13mm and 10.5–14mm, respectively) are the most practical means to distinguish between the two species. The lesser white-toothed shrew is also smaller (head and body length 50–75mm) and lighter (3–7g) than the greater white-toothed shrew (60–90mm and 4.5–14.5g, respectively).

Distribution: This is a widely distributed species. In Europe, it occupies a scattered distribution throughout western and south-eastern France, northern and south-western Spain, Italy and much of eastern Europe. Absent from the mainland of Great Britain and Ireland but present on the Scilly Isles, Jersey, Sark, Ouessant and Yeu. Also present on some of the Balearic Islands, Sardinia and Sicily. This species is listed as 'Endangered' in the state of Mecklenburg-Vorpommern, Germany (Ingelög *et al.*, 1993). *C. suaveolens* is also found in North Africa (Morocco and Algeria) and ranges widely eastwards to Japan – only Tsushima Island (Abe, pers. comm.) – Korea, China and Taiwan.

Habitat: Temperate woodlands and steppe. Similar to *C. russula*.

Ecology and behaviour: The lesser white-toothed shrew is essentially a solitary species, not nearly as pugnacious as *Sorex*. Home ranges of individuals overlap so the species is probably not highly territorial (Churchfield, 1991).

The breeding season extends from March to September. Females have post-partum oestrus which permits lactation and pregnancy to occur simultaneously. May produce 3–4 litters per annum in the wild. In captivity, one litter each month has been recorded. Some breed even in the year of birth (Rood, 1965). The life span of *C. suaveolens* in the wild may be as long as 18 months (in Poland) (Huminski and Wojcik-Migala, 1967).

Crocidura susiana

Taxonomy: *Crocidura susiana* Redding and Lay 1978.

IUCN Category of Threat: Endangered (B1 and 2c).

Distribution: This species is only known from the vicinity of Dezful (Khuzistan Province, south-west Iran), but its range may be more extensive. The taxonomic status of this species is uncertain and further investigations are required to determine its status and precise range.

Crocidura tenuis

Taxonomy: *Sorex tenuis* Müller 1840. This species is known from only two damaged specimens and may belong with *C. fuliginosa*; see Jenkins (1982). More recently, Ruedi (1994) has suggested that *C. tenuis* is a proper species based on karyological data of one specimen.

IUCN Category of Threat: Vulnerable (B1 and 2c).

Description: Although only known from two specimens, this species resembles *C. fuliginosa* in size, with its robust teeth and short tail, but lacks the caudal vibrissae, as does *C. orientalis*.

Distribution: This species is only found on the island of Timor within the Indonesian Archipelago.

Crocidura whitakeri

Taxonomy: *Crocidura whitakeri* de Winton 1898.

IUCN Category of Threat: Lower Risk (subcategory Least Concern).

Distribution: Apart from one record from coastal Egypt, this species is known from the Atlantic and Mediterranean parts of Morocco, Algeria and Tunisia.

Crocidura zarudnyi

Taxonomy: *Crocidura zarudnyi* Ognev 1928. Previously described as *C. pergrisea* but given specific rank following Hassinger (1973).

IUCN Category of Threat: Lower Risk (subcategory Least Concern).

Distribution: This species has been recorded in western Pakistan and Afghanistan, and also in south-east Iran.

Habitat: Little is known about the preferred habitat of this species. Recorded specimens have been collected through a wide range of altitudes ranging from 100–2250m.

Crocidura zimmermanni

Taxonomy: *Crocidura zimmermanni* Wettstein 1953. Formerly included in *C. russula*; but see Hutterer (1993).

IUCN Category of Threat: Vulnerable (B1 and 2c).

Distribution: This species in only known from the highland regions of the island of Crete.

Ecology and behaviour: The behaviour of this species in the wild is unknown.

GENUS DIPLOMESODON

A distinct, monospecific genus characterised by the presence of only two upper unicuspid teeth on each side and, more readily, by the uniquely patterned pelage.

Piebald shrew (Diplomesodon pulchellum)

Taxonomy: *Diplomesodon pulchellus* Lichtenstein 1823.

IUCN Category of Threat: Lower Risk (subcategory Least Concern).

Description: As its name suggests, one of the distinguishing features about this species is the striking pattern of coloration. Upper parts are greyish with an elongated oval patch of white in the middle of the back. The under parts, feet and tail are also white. All the hairs are grey at the base. The fur is soft and fine and the vibrissae are long and numerous, resembling the whiskers of desert-dwelling rodents more than the vibrissae of other shrews. The palm and digits of the forelimbs are

One of the most attractive and striking of the Eurasian shrews the piebald shrew (*Diplomesodon pulchellum*) is adapted to living in desert conditions. (Photo by Peter Vogel)

fringed on both sides with long, stiff, elastic hairs which increase the surface area of the paws, providing better support on loose sand. As with most desert-dwelling species, the ears are quite large.

Distribution: The Piebald, or Turkestan desert shrew *D. pulchellum* is the only species within this genus, occurring in desert environments within Turkmenistan and southern Kazakhstan (Kirghiz Steppes). It is not found west of the Volga and is apparently confined to the plains of Semirechiva in the east.

Habitat: Piebald shrews inhabit several types of desert, preferring damp sands and apparently avoiding areas of shifting sand.

Ecology and behaviour: These shrews are active throughout the year, sometimes by day but mostly at night. They are usually solitary and frequently change their shelters, using crevices, stacks of dried forage and human dwellings. Piebald shrews are capable of digging in the sand, but probably do not dig their own burrows. In the Volga-Ural sandy areas this species occurs with *Hemiechinus auritus* and *Crocidura suaveolens*, which may offer some degree of feeding competition.

Its diet consists mainly of lizards and insects. In sandy areas of the Volga-Ural region, the breeding season has been recorded from April to August, with an average of five young per litter. Females probably give birth more than once a season.

GENUS FEROCULUS

The genus *Feroculus* contains just a single species which is confined to montane forests of Sri Lanka. Almost nothing is known about this species in the wild.

Kelaart's long-clawed shrew
(*Feroculus feroculus*)

Taxonomy: *Feroculus feroculus* Kelaart 1850. Closely related to the African *Sylviosorex* and to *Suncus*.

IUCN Category of Threat: Endangered (B1 and 2c).

Description: Body length ranges from 106–118mm with a tail length of 56–73mm. One individual weighed 35g. The dorsal coloration is ashy black; underparts are a pale colour. Forefeet are nearly white, hindfeet fleshy grey. Tail is a dusky colour with a few whitish hairs at the tip. The fur is close, soft and short. The tail is covered by scanty and fine hairs with a few long, bristle-like hairs scattered along it. The fore feet are large, with long claws; the hind feet are smaller. Dental formula is i 3/2, c 1/0, pm 2/1, m 3/3 × 2 = 30, which differs from *Solisorex pearsoni*, which is also endemic to Sri Lanka.

Distribution: This single species is restricted to the island of Sri Lanka where it inhabits montane forest at elevations of approximately 1850 to 2150m. Fewer than 10 specimens have been recorded.

Habitat: Montane swamps and marshes around 2000m.

Ecology and behaviour: Very little is known about this species. Its discoverer, Kelaart, thought that it was a water shrew, but later investigators believed it to be a semi-fossorial species.

GENUS *SOLISOREX*

The genus *Solisorex* contains just a single species, *S. pearsoni* which has only been recorded from the central highlands of Sri Lanka.

Pearson's long-clawed shrew
(*Solisorex pearsoni*)

Taxonomy: *Solisorex pearsoni* Thomas 1924.

IUCN Category of Threat: Endangered (B1 and 2c).

Description: This species is larger than *F. feroculus*, with a body length of 125–134mm, and tail length of 59–66mm. The fur is soft, close and fine, although not of the same velvet texture as observed in most shrews. The general colour is dark grey-brown with light tipped hairs that give the coat a glossy sheen. The ventral surface is slightly paler. The feet are brown with the forelimbs being enlarged and bearing exceptionally long claws.

The slender tail is closely haired and lacks the scattered long hairs which are present on *F. feroculus*. The ears are small and fully furred. Teeth are large and heavy and the anterior incisors are well developed. The dental formula of *Solisorex* is i 3/2, c 1/0, pm 1/1, m 3/3 × 2 = 28.

Distribution: The single species in this genus is apparently restricted in distribution to the central highlands of Sri Lanka at elevations of 1100–1850m.

Habitat: 'Virgin forest'.

Ecology and behaviour: The habits of this species are unknown.

GENUS *SUNCUS*

This genus of 16 species is found in Africa, Madagascar, southern Europe and southern Asia eastward to the Philippine Islands and New Guinea. Eleven species are represented in Eurasia. Members of the genus differ widely in their habitat requirements. One species in particular, *S. murinus*, has become accustomed to living in and around human habitations and has been repeatedly and unintentionally introduced by man. This is certainly how they reached Guam where they now inhabit buildings, grassy areas and swamps.

The genus *Suncus* contains some of the largest of the true shrews. It also contains one of the smallest known mammals in the world – *S. etruscus* of the Mediterranean, North and West Africa and Asian regions. In general, members of the genus are solitary and intolerant of one another. Sounds are frequently associate with aggressive behaviour and a wide range of vocalisations have been described (Gould, 1969; Roberts, 1977). In China, *S. murinus* is known as the 'money shrew' because of a fancied resemblance between its rather constant, small chatter and the sound of jingling coins! (Walker, 1991).

Black shrew
(*Suncus ater*)

Taxonomy: *Suncus ater* Medway 1965.

IUCN Category of Threat: Critically Endangered (B1 and 2c).

Description: In appearance, this species, measuring 75mm with a tail length of 57mm, is a dark blackish brown colour, with slightly paler underparts. The long tail bears scattered long, dark hairs at the base.

Distribution: The black shrew is known only from a single specimen trapped on Mount Kinabalu, Borneo (1700m).

Habitat: Montane forest.

Ecology and behaviour: Nothing is known about the ecology or status of this species.

Suncus dayi

Taxonomy: *Suncus dayi* Dobson 1888.

IUCN Category of Threat: Endangered (B1 and 2c).

Distribution: This species has only been recorded from Trichur and Cochin (Kerala State), southern India.

Pygmy white-toothed shrew (*Suncus etruscus*)

Taxonomy: *Suncus etruscus* Savi 1822.

IUCN Category of Threat: Lower Risk (subcategory Least Concern).

Description: This species is one of the smallest known mammals alive today, weighing just 2.5g and measuring some 35–50mm, with a tail length of 25–30mm. Dorsal coloration varies from light greyish-brown to dark brown. This species may also be distinguished by its especially short hind limbs.

Distribution: *S. etruscus* has a wide distribution: in Europe, it is mainly confined to the Mediterranean lowlands, ranging from southern Portugal to Greece and the Near and Middle East. It has also been reported from the Atlantic coastline of France. Elsewhere in Europe, it inhabits the islands of Crete, Corsica, Sardinia, Sicily, Majorca and Corfu. In Asia, it is found throughout India and the Himalayan foothills from Pakistan through Nepal and Bhutan (Hutterer, 1979) to Yunnan (China), Myanmar, Thailand and Malaya.

The pygmy white-toothed shrew (*Suncus etruscus*) is the smallest mammal known. (Photo by Peter Vogel)

A defensive posture adopted by two pygmy white-toothed shrews (*Suncus etruscus*) highlights the intolerance which most soricids display towards conspecifics. (Photo by Peter Vogel)

It also occurs on Sri Lanka, as well as Borneo (if *S. hosei* is conspecific). *S. etruscus* has also been recorded from North Africa (ranging from Tunisia to Morocco), and may occur in Ethiopia and West Africa (Northern Nigeria and Guinea).

Habitat: This widespread species has been found in scrub, grasslands, and gardens, often found under stones and logs. In Asia it has been recorded from tall dipterocarp forest. It has been collected at an altitude of 1000m in Italy and 630m in France.

Ecology and behaviour: Little is known about the ecology of this species in the wild, its size preventing many field studies. It is known to be commensal and its wide distribution has almost certainly been assisted by humans. Several investigations have been conducted under captive conditions (see Fons, 1974, 1975).

Suncus fellowsgordoni

Taxonomy: *Suncus fellowsgordoni* Phillips 1932. Formerly included in *S. etruscus*; but see Hutterer (1993).

IUCN Category of Threat: Endangered (B1 and 2c).

Distribution: This species is known only from the central highlands of Sri Lanka.

Suncus hosei

Taxonomy: *Suncus hosei* Thomas 1893. Regarded as a subspecies of *S. etruscus* by Medway (1977) and Payne *et al.*, (1985). Formerly included *S. etruscus*; but see Hutterer (1993).

IUCN Category of Threat: Vulnerable (B1 and 2c).

Description: Similar in external appearance to *S. etruscus*.

Distribution: Known from Sabah and northern Sarawak, Borneo.

Habitat: Dipterocarp forest.

Ecology and behaviour: No information is available on the behaviour of this species.

Suncus malayanus

Taxonomy: *Suncus malayanus* Kloss 1917. Formerly included in *S. etruscus*, but see Hutterer (1993).

IUCN Category of Threat: Data Deficient.

Distribution: Malayan Peninsula. The type specimen was collected at Bang Nari, Pattani, Peninsular Thailand.

Habitat: Tropical forest

Flores shrew
(Suncus mertensi)

Taxonomy: Suncus mertensi Kock 1974.

IUCN Category of Threat: Critically Endangered (B1 and 2c).

Description: A medium-sized, non-commensal species.

Distribution: This species has only been recorded on Flores Island, Indonesia; it is known only from the type specimen.

Suncus montanus

Taxonomy: *Suncus montanus* Kelaart 1850. One subspecies *S. niger malabaricus* has been tentatively proposed from India (Corbet, 1992).

IUCN Category of Threat: Vulnerable (B1 and 2c).

Distribution: Central and southern Sri Lanka; the Nilgiri and Palni Hills and perhaps other parts of southern India.

Habitat: Mostly montane but has been recorded as low as 150m in the rainforests of southern Sri Lanka. Rarely enters buildings.

House shrew
(Suncus murinus)

Taxonomy: *Suncus murinus* Linnaeus 1766.

IUCN Category of Threat: Lower Risk (subcategory Least Concern).

Description: The entire body and tail are a uniform mid-grey to brownish-grey colour. The tail is thick, especially at the base, narrower at the tip, and is covered with a few long, bristle-like hairs that are thinly scattered. This species emits a strong odour of musk, derived from musk glands that are sometimes visible on each side of the body. Odour is especially noticeable during the breeding season. Usually found in or near houses.

Distribution: This species has an extremely wide distribution, ranging throughout the Oriental region. Much of this has taken place as a result of non-deliberate introductions by man. This species is also found throughout Iran and Arabia to Egypt, in eastern Africa, Madagascar and other islands in the Indian Ocean (Réunion, Comoros) and Pacific Ocean (Guam, etc.) and in southern Japan. It may also be present in New Guinea.

Habitat: This species has been recorded up to 2825m near Darjeeling, West Bengal, but only to 300m in Taiwan (Jameson and Jones, 1977).

Ecology and behaviour: The house shrew is a commensal species, usually found near human habitation, rice fields and grain warehouses. Although almost universally disliked, this species is beneficial to man because its diet consists mostly of harmful insects. House shrews breed throughout the year with each female averaging two litters per year. Litter size is 1–5, usually three.

Anderson's shrew
(Suncus stoliczkanus)

Taxonomy: *Suncus stoliczkanus* Anderson 1877.

IUCN Category of Threat: Lower Risk (subcategory Least Concern).

Description: A medium-sized shrew characterised by comparatively large ears and a pale grey coloration. The tail measures about 50–70% of body length. Yellow fur around throat or pectoral region is not always evident.

Distribution: A widespread species represented in Pakistan, Nepal, India and Bangladesh.

Habitat: Gardens and grassy embankments near water courses (Sind and Punjab regions, India) and under piles of brush wood in forest plantations (Punjab) as well as the base of stone walls in Kathiawar (Roberts, 1977); also desert and arid country (Hutterer, 1993).

Ecology and behaviour: As far as is known, this shrew is largely nocturnal and solitary in habits. Breeding may extend throughout the year.

Suncus zeylanicus

Taxonomy: *Suncus zeylanicus* Phillips 1928.

IUCN Category of Threat: Endangered (B1 and 2c).

Distribution: Central and southern Sri Lanka in moist forest; between 150 and 1000m in Sabaragamuwa and Central Provinces.

Habitat: Primary forest

Sub-family Soricinae

The sub-family Soricinae is represented by seven genera (57 species) in Eurasia, as shown in Table 2.4.

GENUS *ANOUROSOREX*

The genus *Anourosorex* contains just a single species of burrowing shrew which occurs in West and Central China, northern Myanmar, northern Thailand, Assam (India), Bhutan, North Vietnam, Taiwan and, possibly, Laos.

Mole shrew (*Anourosorex squamipes*)

Taxonomy: *Anourosorex squamipes* Milne-Edwards 1872. A number of subspecies have been proposed but it is doubtful if these justify recognition (Corbet, 1992), apart from the insular form *yamashinai* on Taiwan.

IUCN Category of Threat: Lower Risk (subcategory Least Concern).

Description: This burrowing shrew measures some 85–108mm, with a short tail of 9–17mm. Eyes are

Table 2.4. Classification of the Eurasian Soricidae – Sub-family Soricinae (Hutterer, 1993)

Genus	Species	Genus	Species
Anourosorex	A. squamipes	Sorex (cont.)	S. hosonoi
			S. isodon
Blarinella	B. quadraticauda		S. kozlovi
	B. wardi		S. leucogaster
			S. minutissimus
Chimarrogale	C. hantu		S. minutus
	C. himalayica		S. mirabilis
	C. phaeura		S. planiceps
	C. platycephala		S. portenkoi
	C. styani		S. raddei
	C. sumatrana		S. roboratus
			S. sadonis
Nectogale	N. elegans		S. samniticus
			S. satunini
Neomys	N. anomalus		S. shinto
	N. fodiens		S. sinalis
	N. schelkovnikovi		S. thibetanus
			S. tundraensis
Sorex	S. alpinus		S. unguiculatus
	S. araneus		S. volnuchini
	S. asper		
	S. bedfordiae	Soriculus	S. caudatus
	S. buchariensis		S. fumidus
	S. caecutiens		S. hypsiblus
	S. camtschatica		S. lamula
	S. cansulus		S. leucops
	S. coronatus		S. macrurus
	S. cylindricauda		S. nigrescens
	S. daphaenodon		S. parca
	S. excelsus		S. salenskii
	S. gracillimus		S. smithii
	S. granarius		

minute and the ears are concealed in the fur. The fur is soft, dense and velvet in texture, being longest on the rump, forming an elevated tuft. There is often a mucilaginous exudate on these elongated hairs. Upper parts are dark olive-grey with paler underparts. The feet are short, broad, naked and scaled with comparatively long claws. The tail is also scaled and is slightly shorter than the hind foot. The nose is long and pointed.

Distribution: The single species in this genus *A. squamipes*, the mole shrew is found from West and Central China (from Shaanxi and Hubei to Yunnan) up to 3000m; northern Myanmar, northern Thailand, Assam, Bhutan, North Vietnam, and on Taiwan from 300–3000m. It may also occur in Laos.

Habitat: Through much of its range, this forest-dwelling species lives at elevations of 1500–3100m in areas of montane forest. It is semi-fossorial, burrowing in leaf litter and loose topsoil. Syntopic with *Soriculus fumidus* in Taiwan.

Ecology and behaviour: No information is available on the behaviour of this species.

GENUS *BLARINELLA*

The genus *Blarinella* comprises two species of burrowing shrews which are restricted to North Myanmar and parts of southern China. One of these species, the southern short-tailed shrew (*Blarinella wardi*), has only recently been recognised as a distinct species.

Northern short-tailed shrew (*Blarinella quadraticauda*)

Taxonomy: *Blarinella quadraticauda* Milne-Edwards 1872.

IUCN Category of Threat: Lower Risk (subcategory Least Concern).

Description: The body form of *Blarinella* is somewhat modified for burrowing, with a stout body and short, slender tail. Body length is approximately 60–75mm, with a short tail measuring 30–40mm. The upper parts are brownish-grey with a silver or smoky-grey reflection. The dorsal fur is a paler colour. Claws on both fore and hind limbs are large and external bodily appendages are reduced, further suggesting that this is a burrowing species.

Distribution: This species has a restricted range in the mountains and highlands of China, specifically Yunnan, Sichuan, Shaanxi and southern Gansu provinces, at altitudes of 2000–3500m (Hoffmann, 1987). Several isolated populations may occur within this region.

Habitat: This specialised species is only found in montane taiga forest.

Southern short-tailed shrew (*Blarinella wardi*)

Taxonomy: *Blarinella wardi* Thomas 1915. Formerly considered a subspecies of *B. quadraticauda*, but separated from it by Hutterer (1993) on the basis of much smaller size.

IUCN Category of Threat: Lower Risk (subcategory Near Threatened).

Description: A slightly smaller species than *B. quadraticauda* which may be distinguished by the much shorter tail length – about half that of the head and body (Hoffmann, 1987).

Distribution: Northern Myanmar and Yunnan (China).

Habitat: The preferred habitat of this species is unknown; it is probably montane forest.

Ecology and behaviour: No additional information is available on this species at the present time.

GENUS *CHIMARROGALE*

The Asiatic water shrews (Genus *Chimarrogale*) are a predominantly Oriental genus of five (Hoffmann, 1987) or six species (Hutterer, 1993). In appearance, all are relatively large shrews which, as their name suggests, are modified for an aquatic life. In all species, the eyes are small and the reduced ears have a valvular flap which seals the ear when submerged. The feet are fringed with stiff hairs on both lateral edges of the digits. The tail is relatively long. All members of this genus have a swollen area at the bases of the vibrissae. As in other semi-aquatic insectivores their dense fur is water repellent and considerable time must be spent grooming to ensure that this is maintained in good condition.

These shrews usually inhabit streams in mountain forests at altitudes of up to 3300m. They are apparently able to swim well under water and are occasionally caught in fish traps (Walker, 1991). Their diet consists of insects, aquatic larvae, crustaceans, and possibly small fish.

Malayan water shrew (*Chimarrogale hantu*)

Taxonomy: *Chimarrogale hantu* Harrison 1958. Formerly included in *C. himalayica* by Medway (1977), and considered a subspecies of *C. phaeura* by Hoffmann (1987) and Corbet (1992); but see Hutterer (1993).

IUCN Category of Threat: Critically Endangered (B1 and 2c).

Distribution: This species has only been recorded from the Malay Peninsula, specifically the Ulu Langat Forest Reserve, Selangor.

Himalayan water shrew (*Chimarrogale himalayica*)

Taxonomy: *Crossopus himalayicus* Gray 1842. Body size decreases from the Himalayas through south-west China, towards south-east China and Taiwan.

IUCN Category of Threat: Lower Risk (subcategory Least Concern).

Description: Body colour is dark grey-brown with conspicuous silvery guard hairs on the hindquarters. Feet are fringed with short, rather stiff hairs, a feature unique to this species. Head and body length ranges from 80–135mm, while tail length is 60–126mm. Body weight ranges from 25–40g.

Distribution: This species is widely distributed throughout southern and western China, as well as in the Himalayan region west to Kashmir, northern Myanmar, Laos, North Vietnam and Taiwan (Jones and Mumford, 1971). Its distribution has recently been mapped by Zheng and Wang (1985).

Habitat: This species frequents mountain streams, at altitudes of 800–1500m in the Himalayas.

Ecology and behaviour: No information is available at the present time.

Borneo water shrew (*Chimarrogale phaeura*)

Taxonomy: *Chimarrogale phaeura* Thomas 1898. This species has previously been included in *C. platycephala* (Ellerman and Morrison-Scott, 1951) but has been designated a separate species both by Hoffmann (1987) and Corbet (1992).

IUCN Category of Threat: Endangered (B1 and 2c).

Description: Probably similar in size and appearance to *C. himalayica*.

Distribution: This species is only known from Mt Kinabalu and Mt Trus Madi, northern Borneo.

Habitat: A semi-aquatic species, it lives in mountain streams from an altitude of 460–1700m.

Ecology and behaviour: No information available.

Chimarrogale platycephala

Taxonomy: *C. platycephala* Temminck 1842. This species has frequently been included with *C. himalayica* but is treated here as a separate species, following Hoffmann (1987).

IUCN Category of Threat: Lower Risk (subcategory Least Concern).

Description: Appearance similar to *C. himalayica*. Head and body length ranges from 103–133mm; tail length from 94–105mm.

Distribution: This species has only been reported from Honshu, Shikoku and Kyushu, Japan (Abe, pers. comm.). See also Hoffmann (1987) and Imaizumi (1970).

Habitat: This species usually inhabits the banks of clear streams and the basins of waterfalls in mountain regions, as high as 1500m (Abe, pers. comm.).

Ecology and behaviour: *C. platycephala* feeds mainly on aquatic insects and small fish. They are active throughout the day. No information is available on their social behaviour.

Styan's water shrew (*Chimarrogale styani*)

Taxonomy: *Chimarrogale styani* de Winton 1899.

IUCN Category of Threat: Lower Risk (subcategory Least Concern).

Description: Body colour predominantly grey, lacking the brown colours of other members of this genus. *C. styani* is also slightly smaller than other species, measuring an average of 108mm.

Distribution: This species has been recorded from south-west China in Sichuan, South Shaanxi, south-east

Qinghai, eastern Tibet, West Yunnan and in north-east Myanmar. Its distribution has been mapped by Zheng and Wang (1985).

Habitat: Clear, unpolluted mountain streams at altitudes of 1570–3100m (in Myanmar).

Ecology and behaviour: No information available at the present time.

Sumatra water shrew (*Chimarrogale sumatrana*)

Taxonomy: *Chimarrogale sumatrana* Thomas 1912. Considered a subspecies of *C. phaeura* by Hoffmann (1987) and Corbet (1992); but see Hutterer (1993).

IUCN Category of Threat: Critically Endangered (B1 and 2c).

Distribution: This species has only been recorded from the Padang highlands, West Sumatra.

GENUS *NECTOGALE*

A single species, the elegant water shrew (*Nectogale elegans*), is represented in this Asian genus. A highly distinctive species, it frequents high altitude mountain streams.

Elegant/Tibetan water shrew (*Nectogale elegans*)

Taxonomy: *Nectogale elegans* Milne-Edwards 1870. One variant, *N. sikhimensis* was based on its larger size and brown pelage. The latter, however, is not judged to be a separate species (Hoffmann, 1987; Corbet, 1992).

IUCN Category of Threat: Lower Risk (subcategory Least Concern).

Description: This attractive species was first described to science by the missionary Père Armand David. These shrews have a soft fur which is slate grey on the head, sides and back, and silver-white on the underside. The long tail (89–110mm) is generally black in colour with two lateral fringes of white hairs which come together and continue along the undersurface to the tip. Two other lateral fringes begin at the first third section of the tail and fade out at the terminal third. Finally, a dorsal fringe of stiff white hairs begins at about the beginning of the terminal third and continues to the tip. Altogether, this gives the tail a four-sided appearance at the base, triangular in section in the middle and laterally compressed towards the tip.

Distribution: The web-footed, elegant, or Tibetan water shrew is found only in the mountain streams of south-west China (South Shaanxi, Sichuan, Yunnan), south-east Tibet, northern Myanmar and the Himalayas west to Sikkim and to eastern Nepal (McNeely, pers. comm.).

Habitat: The elegant water shrew inhabits clean mountain streams at altitudes of about 900–2270m.

Ecology and behaviour: This species swims and dives remarkably well and shelters in burrows in stream banks. Its diet is composed of small fish and aquatic invertebrates. No other details are available at present.

GENUS *NEOMYS*

A clearly defined genus confined to the Palaearctic region. Three species are recognised, each distinguished from other shrews by their dark dorsal fur, red-tipped teeth, presence of four pairs of upper unicuspid teeth and smooth, unlobed first lower incisors. All are semi-aquatic species.

Southern water shrew (*Neomys anomalus*)

Taxonomy: *Neomys anomalus* Cabrera 1907.

IUCN Category of Threat: Lower Risk (subcategory Least Concern).

Description: Slightly smaller (head and body length 65–85mm) than *N. fodiens*. Distinguished from *N. fodiens* by the lesser development of the keel of specialised hairs under the tail. Such a keel is either absent or confined to the terminal third of the tail. Underside of the body is also more consistently pale than in *N. fodiens*.

Distribution: A wide but discontinuous range in the mountains of west and central Europe: in Portugal, Spain, the Massif Central of France, Alps, Italy, Germany (west of the Eifel), Belgium, Balkans, the lowlands of eastern Europe east to Crimea and the River Don; perhaps west to Asia Minor (Osborn, 1965) and, perhaps, Iran. In eastern Europe, it is more common on low ground.

Habitat: Similar to *N. fodiens*. Both species may occur together.

Ecology and behaviour: Probably similar to *N. fodiens*, although this species has not been well studied.

The southern water shrew (*Neomys anomalus*) has a discontinuous range in the mountains of west and central Europe. (Photo by Peter Vogel)

Eurasian water shrew (*Neomys fodiens*)

Taxonomy: *Neomys fodiens* Pennant 1771.

IUCN Category of Threat: Lower Risk (subcategory Least Concern).

Description: One of the largest (head and body length 70–90mm; weight 12–18g) and darkest of the European shrews. The upper side is almost black while the underside is very variable in colour – sometimes pale silvery-grey, but usually suffused with brown and occasionally black. The tail has a prominent keel of stiff silvery hairs extending the whole length of the underside. The hind feet have similar fringes.

Distribution: *Neomys fodiens* is found throughout most of Europe except Iceland, Ireland and most of Iberia (apart from the Pyrenees and northern Spain). It also occurs on most Mediterranean islands, and has been recorded from Albania, Greece and Turkey. Further east, it ranges through western Siberia as far as the Yenesei River and Lake Baikal, and south to northern Asia Minor, Tien Shan and north-western Mongolia. In east Siberia it has an apparently disjunct population from north-eastern China, North Korea and Vladivostok to the mouth of the Amur. It has also been recorded from the island of Sakhalin.

Habitat: Rivers, streams and lakes with abundant riparian vegetation, as well as ponds, marshes, water-cress beds and on boulder-strewn sea shores. Occasionally found far from water. Recorded up to 2500m in the Alps.

Ecology and behaviour: Essentially a solitary species outside of the breeding season. In captivity it exhibits a marked territorial behaviour. Home ranges extend from 20–30m^2 on land, to 60–80m^2 (including water surface) along brooks in Germany (Illing *et al.*, 1981). It appears to spend only a brief time (a few months) in one area before moving on, exhibiting an intermittent nomadic existence with frequent shifts of home range (Churchfield, 1984; Pitt, 1945; Shillito, 1963).

Neomys fodiens displays a preference for living along the banks of clear, fast-flowing unpolluted rivers and streams, but is also found by ponds and drainage ditches. The species occasionally occurs far from water (up to 3km) in deciduous woodland, hedgerows and grassland (Churchfield, 1991). In Poland, it has also been found in coniferous woodland (Dehnel, 1950).

The Eurasian water shrew is active by day and night but with an apparent preference for darkness. Peak activity occurs just before dawn and it is least active in the late morning (Crowcroft, 1954). Usually establishes a burrow system which it creates itself. Foraging is conducted in the water and on the land surface, feeding on a wide range of invertebrates, as well as frogs, newts and small fish. This species paralyses large prey with

The Eurasian water shrew (*Neomys fodiens*) is one of the largest and most easily recognised of all shrews. (Photo by Peter Vogel)

Equally at home on land and in water the Eurasian water shrew (*Neomys fodiens*) forages for aquatic snails and insect larvae amongst the detritus on stream beds. (Photo by Peter Vogel)

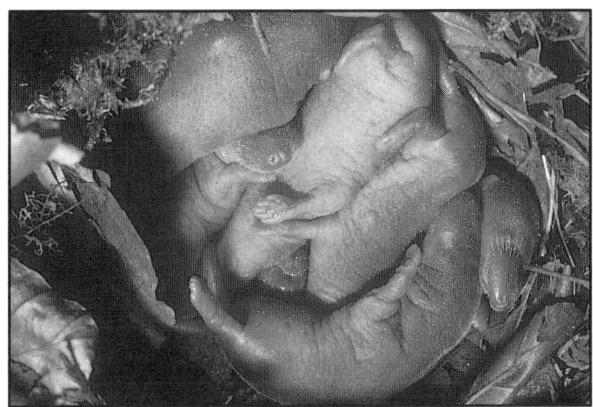

Juvenile Eurasian water shrews (*Neomys fodiens*) are cared for entirely by their mother. (Photo by Peter Vogel)

venomous saliva. In Europe, the diet overlaps considerably with that of *Sorex araneus* and *S. minutus*.

The breeding season extends from April to September, reaching a peak in May and June (Price, 1953). Two litters are generally produced each breeding season.

Transcaucasian water shrew (*Neomys schelkovnikovi*)

Taxonomy: *Neomys schelkovnikovi* Satunin 1913.

IUCN Category of Threat: Lower Risk (subcategory Least Concern).

Description: Dorsal surface velvet brown-black in appearance; ventral parts dark grey. Longer fur than other water shrews. Length of body and head 80mm; tail 62mm. Tip of tail completely white.

Distribution: The only records of this species are from Armenia, Georgia, and Azerbaijan. It may also occur in adjacent Turkey and Iran.

GENUS *SOREX*

The genus *Sorex* ('red-toothed shrews') consists of about 70 species which inhabit the northern hemisphere ranging as far south as Central America in the New World, and to Israel, Asia Minor, Kashmir and northern Myanmar, Thailand and Vietnam in the Old World. Thirty four species are represented in Eurasia. They frequent moist areas, including forests, shrub-grown tracts and tundra. A few species have adapted to living in arid ecosystems.

Alpine shrew (*Sorex alpinus*)

Taxonomy: *Sorex alpinus* Schinz 1837.

IUCN Category of Threat: Lower Risk (subcategory Least Concern).

Description: The alpine shrew is slightly larger than the common shrew *Sorex araneus* (head and body length 60–75mm) and the tail is as long as the head and body combined (it is the longest of the European shrew species). It is a slate grey colour, only slightly lighter on the ventral surface.

Distribution: Mountains of Central Europe to the Pyrenees (isolated populations) and Balkans. Between 600 and 1500m in the Alps, down to 180m in southern Germany.

Habitat: Montane coniferous forests, especially near water. More strictly montane in the south than in the northern part of its range.

Ecology and behaviour: The ecology of this species in the wild is poorly known (see Spitzbergen, 1990). Throughout its range *S. alpinus* overlaps both geographically and ecologically with *S. araneus*, *S. minutus* and *Neomys fodiens*.

Although widely distributed, the ecology of the alpine shrew (*Sorex alpinus*) is poorly known. (Photo by Peter Vogel)

Eurasian common shrew (*Sorex araneus*)

Taxonomy: *Sorex araneus* Linnaeus 1758.

IUCN Category of Threat: Lower Risk (subcategory Least Concern).

Description: This species is recognisable by its tri-coloured coat: upper parts vary from a medium-brown in juveniles to dark brown in adults, with a pale belly. A distinct band of intermediate colour separates the dark and light sections on each flank. The tail is evenly haired.

Head and body length varies from 48–80mm, with tail length ranging from 24–44mm. Weight is 5–14g. The common shrew is noticeably larger than the pygmy shrew (*S. minutus*), but only distinguishable from *S. coronatus* and *S. granarius* by skull measurements and/or chromosome examinations.

Distribution: Europe, including Great Britain and the Pyrenees, but absent from Iberia, most of France and Ireland. This species overlaps slightly with Millet's shrew (*S. coronatus*) in a zone from the Netherlands to Switzerland. It extends eastwards as far as Lake Baikal in all but the dry steppe and desert zones.

Habitat: Found in a wide range of habitats including woodlands, grassland, hedgerows, heath, dunes and scree. May live up to the limits of summer snow line.

Ecology and behaviour: Solitary and aggressive. Young disperse shortly after weaning and individuals of both sexes establish their own home ranges. These are largely exclusive to other shrews of the same species, varying in size from 370–630m^2. Active during day and night with about 10 periods of almost continuous activity (Churchfield, 1991). Juveniles are territorial, and breeding is delayed until their second year. Population densities are highly variable but may range from 42–69/ha in deciduous woodland and grassland during summer months (Crowcroft, 1954; Shillito, 1963; Yalden, 1974). Densities are much lower in winter – 5–27/ha (Churchfield, 1991).

The Eurasian common shrew makes its own surface runways through the ground vegetation, but may also use the subterranean burrows of mice, voles and moles (*Talpa* spp.). They are opportunistic feeders, preying on a wide range of insects, spiders, earthworms and woodlice. Main predators are owls, but stoats, weasels, foxes and cats also prey on shrews. This species constitutes about 5% by weight in the diets of tawny owls in woodland (Southern, 1954) and 6–13% in the diet of barn owls (Buckley and Goldsmith, 1975). This species is especially vulnerable during the juvenile period of dispersal.

Tien Shan shrew
(*Sorex asper*)

Taxonomy: *Sorex asper* Thomas 1914.

IUCN Category of Threat: Lower Risk (subcategory Least Concern).

Description: Moderate-sized shrew: head and body length measures 65mm; tail 37mm. Uniform brown colour in summer; greyer in winter.

Distribution: The Tien Shan shrew has only been recorded from the western Tien Shan mountains, Kyrgyzstan, and the adjacent part of Xinjiang Province, China.

Lesser stripe-backed shrew
(*Sorex bedfordiae*)

Taxonomy: *Sorex bedfordiae* Thomas 1911. Three subspecies have been proposed: *S.b. fumeolus*; *S.b. gomphus* and *S.b. nepalensis*.

IUCN Category of Threat: Lower Risk (subcategory Least Concern).

Description: Nepalese specimens of this shrew are relatively large; head and body length 66–71mm, tail length 55–57mm, hind foot 13.3–13.7mm. The body size is similar to the greater stripe-backed shrew (*Sorex cylindricauda*) in China (Abe, pers. comm.), although Hoffmann (1987) reports that *S. bedfordiae* is smaller than *S. cylindricauda* in all cranial measurements. *S. bedfordiae* populations exhibit geographic variation in relative tail length, longer tails being found in the Sichuan population (90–98% of head-body length), where they are potentially sympatric with *S. cylindricauda*. In Gansu and Yunnan (China) and parts of Myanmar, where *S. bedfordiae* is the only striped shrew, relative tail length is less, being 83–86% of head-body length (Hoffmann, 1987).

Distribution: This species has been recorded from South Gansu, Sichuan and West Yunnan (China), North Myanmar and Nepal at altitudes of 2100–4400m.

Habitat: Montane forest.

Pamir shrew
(*Sorex buchariensis*)

Taxonomy: *Sorex buchariensis* Ognev 1921. Considered conspecific with *S. thibetanus* by Dolgov and Hoffmann (1977) and Hoffmann (in press).

IUCN Category of Threat: Lower Risk (subcategory Least Concern).

Distribution: *S. buchariensis* occurs in the Pamir Mountains, Tadzhikistan and western Tibet.

Laxmann's shrew
(*Sorex caecutiens*)

Taxonomy: *Sorex caecutiens* Laxmann 1788.

IUCN Category of Threat: Lower Risk (subcategory Least Concern).

Description: This medium-sized shrew is intermediate in size between *S. minutus* and *S. araneus*, measuring 50–70mm, excluding tail. In appearance *S. caecutiens* is bicoloured, but darker above than *S. minutus*. In juveniles, the tail is especially well tufted.

Distribution: This species occurs from eastern Europe (Sweden and Poland) to eastern Siberia and south to Mongolia and Korea. Also present on the island of Sakhalin (Hutterer, 1993).

Habitat: Coniferous forest, shrublands and tundra.

Ecology and behaviour: Often found with *Sorex unguiculatus* in the forests of Hokkaido, they feed on a wide range of insects, spiders and centipedes. Breeding season ranges from June to October; 4–8 young are produced per litter. The maximum life span in the wild may be as much as 17 months, in Hokkaido (Abe, 1967).

Kamchatka shrew
(*Sorex camtschatica*)

Taxonomy: *Sorex camtschatica* Yudin 1972. Formerly assigned to *S. cinereus*, a North American species by Corbet (1978), but see Hutterer (1993).

IUCN Category of Threat: Lower Risk (subcategory Least Concern).

Distribution: This species occurs only in the southern Kamchatka Peninsula, Russia.

Habitat: Forest and tundra.

Gansu shrew
(*Sorex cansulus*)

Taxonomy: *Sorex cansulus* Thomas 1912. Considered conspecific with *S. caeutiens* by Corbet (1978), but see Hoffmann (1987).

Status & Summary: Critically Endangered (B1 and 2c).

Distribution: This species is only known from its type locality in Gansu Province, China.

Millet's shrew
(*Sorex coronatus*)

Taxonomy: *Sorex coronatus* Millet 1828.

IUCN Category of Threat: Lower Risk (subcategory Least Concern).

Description: Similar in appearance, but slightly smaller than *S. araneus*. Small morphological differences have been identified but the most reliable means of identification is by chromosomal examination (see Churchfield, 1991).

Distribution: Northern Spain to the Netherlands, southwestern Germany, Switzerland, overlapping slightly with *S. araneus*. Replaces *S. araneus* on Jersey (Channel Islands). Absent from Great Britain.

Habitat: As for *S. araneus*. On Jersey, *S. coronatus* is found in coastal habitats of sand dunes, heath and scrub as well as inland in deciduous woodland, hedgerows and gardens (Godfrey, 1978).

Ecology and behaviour: Similar to *S. araneus*.

Greater stripe-backed shrew
(*Sorex cylindricauda*)

Taxonomy: *Sorex cylindricauda* Milne-Edwards 1872.

IUCN Category of Threat: Endangered (B1 and 2c).

Distribution: This species has only been recorded from Baoxing, North Sichuan, China at an altitude of about 3000m.

Habitat: Montane forest

Large-toothed Siberian shrew
(*Sorex daphaenodon*)

Taxonomy: *Sorex daphaenodon* Thomas 1907.

IUCN Category of Threat: Lower Risk (subcategory Least Concern).

Distribution: Siberia from just east of the Urals to the river Kolyma. Also Kamchatka and the islands of Sakhalin and Paramushir (North Kuriles); northern Mongolia, Manchuria and inner Mongolia (Wang, 1959).

Habitat: Coniferous forest and tundra.

Sorex excelsus

Taxonomy: *Sorex excelsus* Allen 1923. Considered conspecific with *S. asper* by Corbet (1978), but see Hoffmann (1987).

IUCN Category of Threat: Data Deficient.

Distribution: This species occurs in Yunnan and Sichuan Provinces, China, and possibly Nepal.

Habitat: The preferred habitat of this species is not known.

Slender shrew
(Sorex gracillimus)

Taxonomy: *Sorex minutus gracillimus* Thomas 1907. Previously treated as a subspecies of *S. minutus*, but is given specific rank by all Russian authors because of clear differences.

IUCN Category of Threat: Lower Risk (subcategory Least Concern).

Description: A small shrew with a relatively short tail: head and body measure 49–58mm, tail length if from 40–46mm.

Distribution: Siberia from the southern shore of the sea of Okhotsk to North Korea and probably Manchuria; the islands of Sakhalin, Shantar, Hokkaido and some of the Kuriles (Corbet, 1978). Common in northern and eastern parts of Hokkaido (Abe, pers. comm.).

Habitat: This species occupies a wide range of habitats, from grassland to forest (Abe, pers. comm.).

Ecology and behaviour: The behaviour and precise habitat requirements of this species are poorly known.

Iberian shrew
(Sorex granarius)

Taxonomy: *Sorex araneus granarius* Miller 1910. Previously included in *S. caecutiens* (Ellerman and Morrison-Scott, 1951) but Hausser *et al.*, (1975) have demonstrated that its karyotype distinguishes it from *S. araneus* and *S. caecutiens*, and that it is also recognisable by its short skull.

IUCN Category of Threat: Lower Risk (subcategory Least Concern).

Description: Similar in appearance to *S. araneus*, but slightly smaller and with the muzzle rather short and broader. Cannot be readily identified without chromosomal examination.

Distribution: This species has a limited distribution in the mountains of northern Portugal and Central Spain. It is likely that there is no contact zone with either *S. araneus* or *S. coronatus*.

Habitat: Possibly similar to *S. araneus*.

Ecology and behaviour: Similar to *S. araneus*.

Azumi shrew
(Sorex hosonoi)

Taxonomy: *Sorex hosonoi* Imaizumi 1954.

IUCN Category of Threat: Vulnerable (B1 and 2c).

Description: This is a larger and longer-tailed relative of *S. minutissimus*. Head and body length measure from 54–66mm, with a tail length of 40–51mm.

Distribution: This species is only found as relict patchy populations in montane areas of Central Honshu, Japan.

Habitat: Montane habitat.

Ecology and behaviour: No information available on the behaviour of this species.

Even-toothed shrew
(Sorex isodon)

Taxonomy: *Sorex isodon* Turov 1924. Considered conspecific with *S. sinalis* by Corbet (1978), but see Hoffmann (1987).

IUCN Category of Threat: Lower Risk (subcategory Least Concern).

Description: A large species with a drab coloured underside, lighter than the dorsal coloration. A wide brain case and narrow rostrum are often distinguishing features.

Distribution: South-east Norway and Finland through Siberia to the Pacific coast; Kamchatka; Sakhalin Island; Kurile Islands; probably also north-east China and Korea (Hutterer, 1993).

Kozlov's shrew
(Sorex kozlovi)

Taxonomy: *Sorex kozlovi* Stroganov 1952.

IUCN Category of Threat: Critically Endangered (B1 and 2c).

Distribution: This species is only known from the type locality at Dze-Chyu (Zi Qu) River (Tibet), a tributary of the Mekong River.

Habitat: The preferred habitat of this species is not known.

Paramushir shrew
(Sorex leucogaster)

Taxonomy: *Sorex leucogaster* Kuroda 1933. Formerly placed in *S. cinerus* or *S. gracillimus* (Corbet, 1978), but see Pavlinov and Rossolimo (1987); includes *S. beringianus* Yudin 1967.

IUCN Category of Threat: Vulnerable (B1 and 2c).

Distribution: This species is probably confined to Paramushir Island, south of the Kamchatka Peninsula (Russia).

Habitat: The preferred habitat of this species is not known.

Least shrew
(Sorex minutissimus)

Taxonomy: *Sorex minutissimus* Zimmermann 1780.

IUCN Category of Threat: Lower Risk (subcategory Least Concern).

Description: This species is recognisable by its diminutive size (1.6–2.5g – comparable to *Suncus etruscus*, although their ranges are separate).

Distribution: From Norway, Sweden and Estonia to East Siberia, Mongolia, China and South Korea. Small populations occur on Hokkaido (Abe, pers. comm.). It is also found on the island of Sakhalin. This species has not been recorded from many sites and is probably more widely distributed than realised.

Habitat: Wet coniferous forest; edge of moors and shrublands. This species is widespread in the Palaearctic taiga zone.

Ecology and behaviour: Little is known of the behaviour or ecology of this species.

Eurasian pygmy shrew
(Sorex minutus)

Taxonomy: *Sorex minutus* Linnaeus 1766.

IUCN Category of Threat: Lower Risk (subcategory Least Concern).

Description: Distinguished from *S. araneus* by its much smaller size (head and body length 40–60mm; tail length 32–46mm) and weight (2.4–6.1g). Pygmy shrews are bicoloured in appearance – brown above and with a dull white ventral pelage. They lack the contrasting coloured flanks of *S. araneus*.

Distribution: This species is found throughout western Europe from Central Spain and the whole of Scandinavia through western Siberia as far as the Yenesei River and Lake Baikal, and south through the mountains of Central Asia, perhaps to Nepal.

Habitat: Often found in the same habitats as *S. araneus* but able to tolerate sparser ground cover.

Ecology and behaviour: Solitary and aggressive towards others of the same species. Territorial behaviour much as in *S. araneus*. Territories of immature animals are largely mutually exclusive but strict territoriality is abandoned at sexual maturity, particularly by males as they search for mates. Home range size varies from 520–1860m^2, depending on season and habitat. Pygmy shrews are active during day and night, spending more time on the surface than underground, unlike other shrews.

As with *S. araneus*, pygmy shrews generally overwinter as immature animals (at least in Europe), maturing the following March and April and breeding from April to October. Population density ranges from 4–11/ha, depending on season and habitat (Churchfield, 1991).

Ussuri shrew
(Sorex mirabilis)

Taxonomy: *Sorex mirabilis* Ognev 1937.

IUCN Category of Threat: Lower Risk (subcategory Least Concern).

Distribution: This species has been recorded from the Ussuri region of eastern Siberia, as well as from North Korea and north-eastern China.

Habitat: The preferred habitat of this species is not known.

Kashmir shrew
(Sorex planiceps)

Taxonomy: *Sorex planiceps* Miller 1911. Considered conspecific with *S. thibeanus* by Dolgov and Hoffmann (1977) and Hoffmann (in press); but see Hutterer (1979).

IUCN Category of Threat: Lower Risk (subcategory Least Concern).

Distribution: Kashmir (India) and North Pakistan.

Chukotka shrew
(*Sorex portenkoi*)

Taxonomy: *Sorex portenkoi* Stroganov 1956. Considered conspecific with *S. cinereus* (Yudin, 1972; Okhotina, 1977) or *S. ugyunak* (Ivanitskaya and Kozlovskii, 1985), but see Zaitsev (1988) and van Zyll de Jong (1991).

IUCN Category of Threat: Lower Risk (subcategory Least Concern).

Distribution: This species has been recorded from north-east Siberia. The type specimen was collected from Koryaksk, North Kamchatka.

Radde's shrew
(*Sorex raddei*)

Taxonomy: *Sorex raddei* Satunin 1895. Includes *S. caucasicus* Satunin 1913.

IUCN Category of Threat: Lower Risk (subcategory Least Concern).

Distribution: *Sorex raddei* has been recorded from the Trans-Caucasus and north-eastern Turkey. It is sympatric with *S. caucasicus*.

Habitat: Damp forest sites on shores of rivers and lakes, covered by dense grassy vegetation (Ognev, 1962).

Flat-skulled shrew
(*Sorex roboratus*)

Taxonomy: *Sorex roboratus* Hollister 1913. Hoffmann (1985) showed that *S. vir* Allen 1914 was a junior synonym.

IUCN Category of Threat: Lower Risk (subcategory Least Concern).

Description: Uniform colour. *S. roboratus* varies geographically in size and relative proportions (Hoffmann, 1985). *Sorex r. roboratus* from the Altai Mountains and the surrounding region is the largest, with a broader, larger skull. Head and body size ranges from 58–80mm, with a tail length of 35–41mm (Hoffmann, 1985).

Distribution: This species has been recorded from eastern Siberia, west as far as the River Ob and south to the Altai, North Mongolia, Lake Baikal and Vladivostok (Dolgov, 1967).

Habitat: The preferred habitat of this species is not known. The holotype was trapped in dense *Pinus cembra* forest (Hoffmann, 1985).

Ecology and behaviour: No information available.

Sado shrew
(*Sorex sadonis*)

Taxonomy: *Sorex sadonis* Yoshiyuki and Imaizumi 1986.

IUCN Category of Threat: Endangered (B1 and 2c).

Distribution: *Sorex sadonis* is only known from Sado Island, Japan.

Appenine shrew
(*Sorex samniticus*)

Taxonomy: *Sorex samniticus* Altobello 1926. Provisionally considered a subspecies of *S. araneus* by Corbet (1978); but separated as distinct by Graf *et al.*, (1979) and Hausser (1990).

IUCN Category of Threat: Lower Risk (subcategory Least Concern).

Description: Similar to *S. araneus*, but the tail of *S. samniticus* is distinctly shorter (usually less than 40mm).

Distribution: This species is only found in Italy, overlapping with *S. araneus* in the Appenines.

Habitat: This species has been recorded up to an altitude of 1160m where it co-exists with *S. araneus*, but also occurs on lower ground in the absence of *S. araneus*.

Caucasian shrew
(*Sorex satunini*)

Taxonomy: *Sorex satunini* Ognev 1922. Formerly called *S. caucasicus* Satunin, which has been shown to be a junior synonym of *S. raddei* (Pavlinov and Rossolimo, 1987).

IUCN Category of Threat: Lower Risk (subcategory Least Concern).

Distribution: This species is found in northern Turkey and ranges from the Central Caucasus to the Azov sea.

Habitat: The preferred habitat of this species is not known.

Shinto shrew
(Sorex shinto)

Taxonomy: *Sorex shinto* Thomas 1905. Considered conspecific with *S. caecutiens* by Abe (1967) and Corbet (1978), but see Pavlinov and Rossolimo (1987) and George (1988).

IUCN Category of Threat: Lower Risk (subcategory Least Concern).

Description: Length of body and head 58–69mm; tail very long – 48–50mm. Back a uniform brownish white with a light olive-rust hue. Flanks lighter than back and gradually changing to dull grey-brown on abdomen.

Distribution: This species is known from Honshu, Shikoku and Hokkaido, Japan.

Dusky shrew
(Sorex sinalis)

Taxonomy: *Sorex sinalis* Thomas 1912. Considered conspecific with *S. isodon* by Corbet (1978); but see Hoffmann (1987).

IUCN Category of Threat: Vulnerable (B1 and 2c).

Description: Similar in size to *S. araneus*, this species is more uniform in colour between dorsal and ventral surfaces and also more grey (see also Hoffmann, 1987). Long tail (55mm) in relation to body length (70mm).

Distribution: Known only from the mountains of west-central China; type specimens collected south-east of Fengsiangfu, Shaanxi and southern Gansu.

Habitat: Mainly moist montane forest.

Tibetan shrew
(Sorex thibetanus)

Taxonomy: *Sorex thibetanus* Kastchenko 1905. Thought to include *S. bucharensis*, *S. kozlovi* and *S. planiceps* by Dolgov and Hoffmann (1987) and Hoffmann (in press); but see Hutterer (1979, 1993).

IUCN Category of Threat: Lower Risk (subcategory Least Concern).

Distribution: This species has been recorded from the Himalayas and north-east Tibet.

Tundra shrew
(Sorex tundrensis)

Taxonomy: *Sorex tundrensis* Merriam 1900. This species was formerly considered as conspecific with *S. arcticus*; but see Youngman (1975) and Junge *et al.*, (1983).

IUCN Category of Threat: Lower Risk (subcategory Least Concern).

Description: Similar in external appearance to *S. araneus*. See Junge *et al.*, (loc.cit.) for detailed discussion.

Distribution: This species is found throughout Siberia (in both taiga and tundra ecosystems) from the Urals to Vladivostok and Anadyr. It is also found in North America from Alaska to the Yukon Territory. It is restricted to a northern distribution, the limits of which are far from clear.

Habitat: The preferred habitat of this species is mixed ground vegetation in well-drained patches of forest and tundra.

Ecology and behaviour: Little information exists on this species in Eurasia. In Canada, data indicate a high reproductive potential, almost certainly an adaptation to Arctic conditions (van Zyll de Jong, 1983). Insects, earthworms and floral parts of small grasses have been identified from the digestive tracts of specimens from Alaska (Quay, 1951).

Long-clawed shrew
(Sorex unguiculatus)

Taxonomy: *Sorex unguiculatus* Dobson 1890.

IUCN Category of Threat: Lower Risk (subcategory Least Concern).

Description: A relatively large shrew with a head and body length of 54–97mm and a tail length of 40–53mm. Its feet and claws are also large.

Distribution: This species has been recorded along the Pacific coastline of Siberia from Vladivostok to the Amur, as well as from the islands of Sakhalin and Hokkaido.

Habitat: This shrew occurs in a wide range of habitats, ranging from wet grasslands to montane forests. It is most common in grasslands and areas of shrub.

Ecology and behaviour: Semi-fossorial in habit, feeding mainly on insects and small earthworms (Abe, 1967;

Abe, pers. comm.). Solitary and territorial, apart from adult males during the breeding season. Active during day and night. A female produces 3–7 young from April to September. A maximum life span of 18 months has been recorded for a free-living shrew from Hokkaido (Abe, 1967).

Ukrainian shrew (*Sorex volnuchini*)

Taxonomy: *Sorex volnuchini* Ognev 1922. Considered a subspecies of *S. minutus* by Corbet (1978), but see Kozlovskii (1973) and Sokolov and Tembotov (1989).

IUCN Category of Threat: Lower Risk (subcategory Least Concern).

Distribution: South Ukraine and Caucasus; possibly Turkey and North Iran.

GENUS *SORICULUS*

This genus, comprising 10 species (Hoffmann, 1986) is represented in Bhutan, northern India, Sikkim, Nepal, China, northern Myanmar, Tonkin (Vietnam) and Taiwan. These species inhabit mainly damp areas in forest but may also frequent thickets. The length of the body varies from about 53–96mm and the length of the tail is 38–120mm. The coloration is reddish-brown, dark brown, greyish or blackish; the under parts are usually somewhat paler. The fur is dense and soft.

Hodgson's brown-toothed shrew (*Soriculus caudatus*)

Taxonomy: *Sorex caudatus* Horsfield 1851. Corbet (1992) notes that slight variation occurs in this species; two subspecies have been proposed: *S.c. umbrinus* (a small, dark form) from Yunnan and *S.c. soluensis* (a small form with a long rostrum) from East Nepal.

IUCN Category of Threat: Lower Risk (subcategory Least Concern).

Description: Head and body length ranges from 61–69mm, with a tail length of 48–56mm.

Distribution: Kashmir to North Myanmar; also known from Yunnan and Sichuan (China) at altitudes of 1800–3600m.

Habitat: A common shrew at the edges of rhododendron and coniferous forests; also found on alpine meadows in Central Nepal (Abe, pers. comm.).

Ecology and behaviour: The ecology of this species in the wild is poorly known.

Taiwan brown-toothed shrew (*Soriculus fumidus*)

Taxonomy: *Soriculus fumidus* Thomas 1913. The form *sodalis* has been considered a synonym, but recent specimens suggest that it is probably a distinct species (Corbet, 1992; A. Wu, pers. comm. to R. Hoffmann).

IUCN Category of Threat: Lower Risk (subcategory Least Concern).

Description: This species is larger than *S. caudatus* with a relatively shorter tail. Morphological identification may be based on its short, narrow rostrum, and a mandible with long angular and coronoid processes.

Distribution: This species is endemic to Taiwan.

Habitat: According to Jameson and Jones (1977), this species is widely distributed in the temperate montane forests of Taiwan, extending upwards into the dwarf bamboo zone. It has been recorded at an altitude of 1000–3200m.

Ecology and behaviour: No information available.

De Winton's shrew (*Soriculus hypsibius*)

Taxonomy: *Soriculus hypsibius* de Winton 1899.

IUCN Category of Threat: Lower Risk (subcategory Least Concern).

Description: This species is only slightly larger than *S. caudatus* with a tail length usually less than that of the head and body combined. Morphological differences are reported in Hoffmann (1986).

Distribution: This species is restricted to two disjunct areas in China – Sichuan and South Shaanxi (Qinling Shan) – and Hebei.

Habitat: Montane forest.

Ecology and behaviour: The ecology of this species in the wild is not known. Hoffmann (loc. cit.) points out that the ecological relationships of this species with *S. caudatus* would be of interest given their similar body sizes and proportions. *Soriculus hypsibius* appears not to occur at localities where *S. caudatus* is found, but this needs verification.

Soriculus lamula

Taxonomy: *Soriculus lamula* Thomas 1912. The Yunnan form, *parva* is slightly smaller and darker than those in Sichuan and Gansu. Considered a subspecies of *S. hypsibius* by Corbet (1978), but see Hoffmann (1986).

IUCN Category of Threat: Lower Risk (subcategory Least Concern).

Description: A small long-tailed shrew which is not well described in literature.

Distribution: This species has been recorded from north-west Yunnan, Central Sichuan and South Gansu, China at about 2000–3000m. A single specimen has also been recorded from Fujian (south-east China). *S. lamula* is sympatric with *S. hypsibius* in several parts of Sichuan (Hoffmann, 1986).

Habitat: The preferred habitat of this species is unknown.

Ecology and behaviour: No information available. The ecological relationships between *S. lamula*, *S. hypsibius* and *S. caudatus* are not known, but should be investigated.

Indian long-tailed shrew (*Soriculus leucops*)

Taxonomy: *Soriculus leucops* Horsfield 1855. Includes *baileyi* and *gruberi*, considered distinct by some authors.

IUCN Category of Threat: Lower Risk (subcategory Least Concern).

Description: A long-tailed shrew with a head and body length of 66mm and a tail length of about 86mm.

Distribution: This species has been recorded from Central Nepal to North Myanmar, south-west Yunnan (China) and North Vietnam at altitudes of up to 2900m (in Myanmar).

Habitat: In Nepal this species occurs in evergreen broadleaved forests at the lower temperate zone.

Ecology and behaviour: Unknown.

Arboreal brown-toothed shrew (*Soriculus macrurus*)

Taxonomy: *Soriculus macrurus* Blanford 1888. Formerly confused with *S. leucops*, but shown to be distinct by Hoffmann (1986).

IUCN Category of Threat: Lower Risk (subcategory Least Concern).

Description: A relatively large shrew with a head and body length of about 82mm and tail length of 76mm.

Distribution: The range of this species extends from Central Nepal to Sikkim, from north-west Myanmar and West and South Yunnan (China) to Sichuan (China) and North Vietnam.

Habitat: An uncommon species which, in Nepal, is found in wet habitats with dwarf bamboo, scrub and grasses in the temperate zone (Abe, pers. comm.).

Ecology and behaviour: No information available.

Sikkim large-clawed shrew (*Soriculus nigrescens*)

Taxonomy: *Soriculus nigrescens* Gray 1842. The following subspecies (and ranges) have been suggested: *S.n. nigrescens* (Himalayas); *S.n. pahari* (Gnatong, Sikkim); *S.n. caurinus* (Khati, Kuman, India); *S.n. centralis* (Bouzini, Nepal) and *S.n. radulus* (Myanmar) (Corbet, 1992).

IUCN Category of Threat: Lower Risk (subcategory Least Concern).

Description: A large shrew with a relatively short tail; head and body length ranges from 91–106mm, tail length from 35–47mm. Large feet and claws.

Distribution: The range of *S. nigrescens* extends through the Himalayas, from Kumaon (northern India) east through Nepal and Bhutan to Tibet and North Myanmar, at altitudes of 1560–4300m.

Habitat: Broadleaf and coniferous forests.

Ecology and behaviour: A semi-fossorial species feeding on insects and earthworms obtained from the leaf litter and humus layer. In Nepal, females with 5–9 embryos have been found during the breeding season – April to June (Abe, 1971). This species is often syntopic with other shrews (Abe, pers. comm.).

Soriculus parca

Taxonomy: *Soriculus parca* Allen 1923. Considered a subspecies of *S. salenskii* by Lekagul and McNeely (1977), but see Hoffmann (1986). Includes *lowei* and *furva* (Hoffmann, 1986)

IUCN Category of Threat: Lower Risk (subcategory Least Concern).

Description: A medium-sized long-tailed shrew, with a head-body length of 66–74mm and a tail length of 74–95mm.

Distribution: Southern Sichuan to north-west Yunnan (China), north-east Myanmar, northern Thailand and North Vietnam at altitudes up to 2700m.

Habitat: Ranges from 1220–2750m; reported from under rocks and logs, frequently near streams.

Ecology and behaviour: Unknown.

Salenski's shrew (*Soriculus salenskii*)

Taxonomy: *Soriculus salenskii* Kastschenko 1907.

IUCN Category of Threat: Critically Endangered (B1 and 2c).

Description: Fine soft fur, black at the base with grey tips. Darker coloration on the rump, with paler underside. Ears are blackish-brown. Long facial vibrissae.

Distribution: This species is only known from the type locality in northern Sichuan, China.

Smith's shrew (*Soriculus smithi*)

Taxonomy: *Soriculus smithii* Thomas 1911. Considered a subspecies of *S. salenskii* by Lekagul and McNeely (1977), but see Hoffmann (1986).

IUCN Category of Threat: Lower Risk (subcategory Least Concern).

Description: A large, long-tailed shrew with a head-body length range of 72–96mm and a tail length of 92–108mm.

Distribution: Central Sichuan and Qinling Mountains, South Shaanxi, China. Possibly conspecific with the little known *Soriculus salenskii*.

Habitat: The preferred habitat of this species is not known.

Ecology and behaviour: No information available.

2.3.3 Family Talpidae: Moles and Desmans

This family of 17 genera and 42 species occurs widely throughout the Old and New World. In Eurasia, 12 genera (35 species) are represented (Table 2.5) with representatives distributed northwards to about 63°N, south to the Mediterranean and east to Japan. Remaining members of the family are widespread in the New World, ranging from southern Canada to northern Mexico.

Sub-family Desmaninae

GENUS *DESMANA*

The genus *Desmana* consists of a single, highly specialised species found only in eastern Europe.

Russian desman (*Desmana moschata*)

Taxonomy: *Desmana moschata* Linnaeus 1758.

IUCN Category of Threat: Vulnerable (B1 and 2c). This species is included in the Red Data book for Russia, Ukraine, Belarus and Kazakhstan.

Description: An aquatic insectivore of still and slow-moving water. Body length ranges from 180–215mm, tail length from 170–215mm. Easily distinguished by the elongated proboscis which is grooved above and below; it is highly flexible and used for exploration. The tail is laterally compressed and is largest at the base where there are special scent glands that give the animal a distinctive, musky odour. Rings of scales encircle the tail, with a few hairs growing between them. The hind feet are webbed to the tips of the toes and along the edges of the feet there are fringes of stiff hairs that further increase the surface area for swimming. The forefeet are only partially webbed and also have hair-fringed edges. The pelage consists of soft, dense underfur which is interspersed with longer, coarser guard hairs. The colour above is a rich reddish-brown, shading to ash-grey beneath.

Distribution: The single species in this genus is found in Russia, Belarus, the Ukraine and Kazakhstan. Desman populations are reported from the basins of the following rivers: Volga (23,000 animals), Don (about 12,000 animals), Dneiper (2000–3000), Ural (2000), and the Uj and Tobal rivers (2500 animals). This species has been the subject of previous re-introduction efforts (see below and Appendix I). There are now five nature reserves and 80 refuges for desmans (Khakhin, 1993).

Table 2.5. Classification of the Talpidae (Hutterer, 1993)

Genus	Species	Genus	Species
Sub-family Desmaninae		*Scapanulus*	*S. oweni*
Desmana	*D. moschata*	*Scaptochirus*	*S. moschatus*
Galemys	*G. pyrenaicus*	*Scaptonyx*	*S. fusicaudus*
Sub-family Talpinae		*Talpa*	*T. altaica*
			T. caeca
Euroscaptor	*E. grandis*		*T. caucasica*
	E. klossi		*T. europaea*
	E. longirostris		*T. levantis*
	E. micrura		*T. occidentalis*
	E. mizura		*T. romana*
	E. parvidens		*T. stankovici*
			T. streeti
Mogera	*M. etigo*		
	M. insularis	*Urotrichus*	*U. pilirostris*
	M. kobeae		*U. talpoides*
	M. minor		
	M. robusta	**Sub-family Uropsilinae**	
	M. tokudae		
	M. wogura	*Uropsilus*	*U. andersoni*
Nesoscaptor	*N. uchidai*		*U. gracilis*
			U. investigator
Parascaptor	*P. leucura*		*U. soricipes*

Habitat: Slow-flowing rivers, lakes, ponds, canals and marshland. Permanent supply of freshwater essential. Occasionally found in brackish waters. Burrows are created in banks, usually under vegetation, and are used for shelter and breeding.

Ecology and behaviour: A semi-aquatic species which feeds on insects, crustaceans, molluscs, fish, amphibians from freshwater streams and lakes. A total of 90 food items have been recorded in its diet (Onufrienja and Onufrienja, 1993). A simple burrow system is constructed with the single opening beneath the water level, leading up and away from the water to a large, single chamber. The social organization of this species is unclear, but it appears to be a social species; as many as eight animals have been discovered in a single den. Onufrienja and Onufrienja (1993) reporting on 54 years of research from the Oka State Reserve, mention that two animals are usually found at a nest site, occasionally three or four. Largely nocturnal, but also seen during daylight. It appears to be somewhat nomadic. Such records may originate from its tendencies to migrate when rising flood waters threaten the nest. When flood waters are high, desmans may nest in trees and bushes. In addition, however, desmans may move from seasonally shallow lakes where they breed, to deeper lakes to over-winter. Onufrienja and Onufrienja (1993) report a close correlation between rainfall and overall desman numbers: when water levels are high, desmans are offered good overwintering conditions. When water levels are low, desmans may need to migrate overland to other overwintering sites.

Breeding may take place throughout the year, but most appears to be concentrated in spring and autumn (Onufrienja and Onufrienja, 1993). Litters of 3–5 young have been recorded (Ognev, 1962). The average litter size recorded from Oka Reserve was 3.6 (Onufrienja and Onufrienja, 1993).

Setting a live trap for the Russian desman (*Desmana moschata*). (Photo by Gennady Khakhin)

In contrast to the fast-flowing habitat of the Pyrenean desman (*Galemys pyrenaicus*) the Russian desman (*Desmana moschata*) prefers shallow lakes and standing water bodies. (Photo by Gennady Khakhin)

More at home in the water, the Russian desman (*Desmana moschata*) occasionally comes on to land to forage and disperse. (Photo by Gennady Khakhin)

Air tracks beneath the spring ice lead to the nest of a Russian desman (*Desmana moschata*). (Photo by Gennady Khakhin)

Presence of the Russian desman (*Desmana moschata*) may be detected by the air tracks they leave beneath the ice cover. (Photo by Gennady Khakhin)

A Russian desman (*Desmana moschata*) caught in a special live trap. (Photo by Gennady Khakhin)

Marking experiments have shown that desmans are sedentary animals as long as environmental conditions remain stable. As a rule, desmans use the same burrows for several years. Radiotracking experiments have shown that adult females have territories ranging from 0.34–0.56ha, with 2–6 nest holes.

The Russian desman was relatively abundant until the late 19th century when hunting pressure for its valuable fur eliminated and decimated many populations. From 1817 to 1819, for example, Russia exported 325,500 skins to China; some 100,000 skins were sold on the home market at the Nizhni Novgorod fair in 1836 (Kaplin, 1960). By the turn of the 20th century, however, about 20,000 skins were being processed annually (Grzimek, 1975). Desman hunting was banned in 1929 and attempts were made to breed this species in captivity. As a result of this initiative, a total of 10,000 animals were released in 30 republics and regions of the former USSR (Khakhin, 1993). Some new populations were established. Lack of management and poor protection, together with some licensed hunting of desmans in 1940, meant that the overall number of animals did not increase. Hunting was again banned in 1957. The number of free-living desmans has continued to decline; estimates suggest that the overall population is now about 40,000–50,000 animals.

Apart from hunting, the main reasons for the decline of this species, both in terms of overall number and distribution, have been water pollution, creation of impoundments, drainage, clearance of riparian vegetation, entanglement in fishing nets and competition for breeding sites with introduced nutria and muskrats.

This species has been the subject of much investigation; one ongoing study at Oka State Reserve, is now entering its 55th year (see Onufrienja and Onufrienja, 1993).

GENUS *GALEMYS*

The genus *Galemys* consists of a single species, a semi-aquatic insectivore which is restricted to the Pyrenees Mountains, northern and Central Spain and northern Portugal.

Pyrenean desman (*Galemys pyrenaicus*)

Taxonomy: *Galemys pyrenaica* E. Geoffroy 1811. One subspecies *G.p. rufulus* has been proposed from Iberia, but supporting evidence of this claim is weak and probably a result of colour variability.

IUCN Category of Threat: Vulnerable (B1 and 2c).

Description: A distinctive species with a body length of approximately 125mm and an elongated cylindrical tail about 140mm long. The short fur is thick, dark brown with a metallic gloss on the back. The pelage appears shiny when the animal is underwater. Ventral coloration is light and silvery. The tail is fringed at the end and slightly flattened vertically. This acts as a rudder when swimming, helping to steer and direct the animal as it swims. The long muzzle is black and almost devoid of fur. The snout is widened at the tip where two partially divided lobes are evident, each with an elongated nostril opening on the upper side. The eyes are small and surrounded by short hair. There is no pinna and the auditory meatus is hidden by the fur. The forelimbs are smaller than the hindlimbs. Its feet are webbed and bear large claws (Palmeirim and Hoffmann, 1983).

Distribution: This species is confined to south-western Europe, notably the French and Spanish Pyrenees, northern Spain and northern Portugal. In France, it occurs along the Aude, Agly, Salat, Aspe, Ossau, Ariège, Audour, Tet and Tech rivers (Richard and Vallette-Viallaird, 1969). A more recent revision of the distribution of this species in France can be found in Bertrand (1993). Its precise distribution is less well known in Spain and Portugal, but is known from along streams in the northern part of the central plain in the surroundings of the Picos de Europe and along the Deva River (Corbet, 1966, Niethammer, 1964). Vericaud (1970) gives the following locations for the Spanish Pyrenees: Sierra de Guarra north of Huesca and in the vicinity of Infiesta (Oviedo) and Burguete (Navarro). Due to its tendency to conceal itself, no estimates of the number of animals can be made within a given area. Detection of faeces deposited on rocks or other objects is not a valid means of assessment since fluctuating water levels quickly remove any signs of evidence.

Habitat: Unlike the Russian desman, the Pyrenean species lives along fast-flowing mountain streams. It is occasionally found in slower-moving water bodies, including canals, lakes and marshes at altitudes of 60–1200m.

Ecology and behaviour: A highly specialised animal adapted to an aquatic environment. The desman forages mainly at night, feeding on a wide range of crustaceans and insect larvae, including mayfly, stonefly and caddisfly larvae.

A semi-aquatic insectivore of fast-flowing streams, the Pyrenean desman (*Galemys pyrenaicus*) spends considerable time grooming itself on rocks on the river bank. (Photo by David Stone)

The tactile and highly sensory snout of the Pyrenean desman (*Galemys pyrenaicus*) assists with the location of prey. (Photo by David Stone)

Its life history and habitat have been studied and there is some information concerning its distribution. Parts of its range are situated within the Parc National des Pyrénées Occidentales and the Parque Nacional de Covadona. It is possible that the species also occurs in the Parque Nacional de Aiguas y Lago de San Mauricio and the Parque Nacional de Ordesa.

This species has declined in recent years because it is bound to a very vulnerable habitat in a restricted area. It is most threatened by water pollution and the construction of hydro-electric plants which fragment the habitat. It has also suffered from direct human persecution either through over-zealous collecting or from fishermen who (mistakenly) perceive it as a threat to fish stocks, especially trout. The full impact of these activities is not known. In Iberia, fears have been expressed on the potential damage to aquatic species following escape of North American mink (*Mustela vison*) from fur farms in northern Iberia.

The ecology of this species has been described in detail by Richard and Vallette-Viallaird (1969) and Stone (1985, 1986, 1987a,b). More recently, the Pyrenean and Russian desmans have been the subject of an international conference, the proceedings of which may be found in Queiroz (1993). (See also Appendix II.)

Sub-family Talpinae

The sub-family Talpinae consists of nine genera with a total of 29 species. All representatives are fossorial, living in a wide range of habitats from lowland fertile plains to montane slopes at elevations up to 3000m. A number of threatened species are recognised in this sub-family: *Euroscaptor mizura, E. parvidens, Mogera etigo, M. tokudae* and *Talpa streeti*. All of these species are known only from single, or a few isolated locations. The single greatest threat to these, and other species of fossorial insectivores, is further habitat destruction, especially reclamation of lowland plains for intensive agriculture, and the clearance of forested hillsides for shifting cultivation. With increasing pressure to produce more food, the threats to these restricted species are likely to increase.

GENUS *EUROSCAPTOR*

This genus consists of six species, all formerly assigned to *Talpa* by Corbet (1978). The current, revised presentation is based on Hutterer (1993).

Euroscaptor grandis

Taxonomy: *Euroscaptor grandis* Miller 1940.

IUCN Category of Threat: Lower Risk (subcategory Least Concern).

Distribution: The range of this species is unclear because of identification problems. Specimens have been recorded from Central Sichuan Province (Omei Shan and Chengdu), as well as western Yunnan, China (Corbet, 1992). Hutterer (1993) also lists this species from north and south Bakbo and Cha-pa (Vietnam).

Euroscaptor klossi

Taxonomy: *Euroscaptor klossi* Thomas 1929.

IUCN Category of Threat: Lower Risk (subcategory Least Concern).

Distribution: Highlands of Thailand, Laos and Peninsular Malaysia.

Ecology and behaviour: According to Cranbrook (1966) *E. klossi* makes shallow runs just under the surface, but may also construct deeper tunnels. Territory size has been suggested at 100–200m^2. *E. klossi* breeds once, rarely twice, each year with a litter size of 2–7. Gestation is about five weeks although delayed implantation has been reported (Lekagul and McNeely, 1977). Probably territorial, as *T. europaea* (see later), but no studies have been reported on this species.

Euroscaptor longirostris

Taxonomy: *Euroscaptor longirostris* Milne-Edwards 1870. Formerly included in *E. micrura* by Corbet (1978); but see Hutterer (1993).

IUCN Category of Threat: Lower Risk (subcategory Least Concern).

Distribution: Sichuan and adjacent areas of South Shaanxi, East Qinghai and West Yunnan (China); North Vietnam.

Habitat: The preferred habitat of this species is not known. Its altitudinal range occurs from 1800–2900m.

Himalayan mole (*Euroscaptor micrura*)

Taxonomy: *Euroscaptor micrura* Hodgson 1841. Several subspecies have been proposed on the basis of fur colour, but because of the tendency of museum specimens to fade (Cranbrook 1962), this has not been clarified. Too few specimens are available to judge geographic variation.

IUCN Category of Threat: Lower Risk (subcategory Least Concern).

Description: Thick dark brown fur with a silver gloss. Pointed snout with long naked nose pad grooved on the upper side. Tail is reduced and concealed by fur. Forelimbs lack fur and are prominent.

Distribution: This species occurs from the eastern Himalayas west probably as far as eastern Nepal. It is also known from isolated localities in Assam and Thailand.

Habitat: Recorded mostly at altitudes of 1000–2000m, but down to 100m in Bengal and Assam.

Japanese mountain mole (*Euroscaptor mizura*)

Taxonomy: *Euroscaptor mizura* Gunther 1880. The subspecies *othai* is "probably a distinct species" (Imaizumi, 1970; see also Yoshiyuki, 1988).

IUCN Category of Threat: Vulnerable (B1 and 2c).

Description: A small, primitive mole with a head and body length of 79–107mm. The tail is relatively long, ranging from 23–26mm. The muzzle has a long triangular naked portion on the upper side. The colour of the pelage varies from brown to black.

Distribution: This species is confined to just a few isolated montane areas on Honshu, Japan.

Habitat: Forest and alpine grassland.

Ecology and behaviour: The diet of the Japanese mountain mole comprises insects, earthworms and centipedes. Other details of its behaviour are unknown.

Euroscaptor parvidens

Taxonomy: *Euroscaptor parvidens* Miller 1940. Formerly included in *E. micrura* or *P. leucura*; but see Hutterer (1993).

IUCN Category of Threat: Critically Endangered (B1 and 2c).

Distribution: This species is only known from the type locality, Di Linh, Vietnam, and Rakho on the Chinese border.

GENUS *MOGERA*

This genus, with seven species recognised here, was formerly included in *Talpa* (Corbet, 1978), but see Hutterer (1993).

Mogera etigo

Taxonomy: *Mogera etigo* Yoshiyuki and Imaizumi 1991. Formerly included in *M. tokudae*, but see Hutterer (1993).

IUCN Category of Threat: Endangered (B1 and 2c).

Distribution: This species has only been recorded from Niigata (formerly Echigo Plain), Honshu, Japan.

Mogera insularis

Taxonomy: *Mogera insularis* Swinhoe 1863. Corbet (1978) and Hutterer (1993) include *M. latouchei* (occurs in south-east China, Hainan) as a subspecies, but see Abe (1988) and Abe *et al.*, (1991).

IUCN Category of Threat: Lower Risk (subcategory Near Threatened).

Distribution: This species has only been recorded from Taiwan and Hainan (south-east China).

Mogera kobeae

Taxonomy: *Mogera kobeae* Thomas 1905. Considered a subspecies of *M. robusta* by Corbet (1978), but see Hutterer (1993).

IUCN Category of Threat: Lower Risk (subcategory Least Concern).

Distribution: Kyushu, Shikoku and the southern region of Honshu, Japan.

Mogera minor

Taxonomy: *Mogera minor* Kuroda 1936. Considered a subspecies of *M. wogera* by Corbet (1978), but see Hutterer (1993).

IUCN Category of Threat: Lower Risk (subcategory Least Concern).

Description: Body size varies considerably according to location, but is generally small compared to *M. wogura*. Head and body length range from 121–159mm; body weight is 48–127g. Forward projecting degree of the upper incisor row is large and the row is V-shaped.

Distribution: Northern half of Honshu, Japan. Several small, relict populations occur in mountainous regions of the southern half of Honshu and in Shikoku.

Habitat: Low-lying wet plains appear to be the preferred habitat of this species, but it also survives in montane forests (Abe, pers. comm.).

Ecology and behaviour: *M. minor* feeds on a wide range of insects, earthworms, leeches, spiders, centipedes and plant seeds. Litter size ranges from 2–6, with an average of 3.6. Maximum longevity is about 3.5 years (Abe, pers. comm.).

Greater mole (*Mogera robusta*)

Taxonomy: *Mogera robusta* Nehring 1891. Includes *M. coreana;* formerly included *kobeae* and *tokudae* (Corbet, 1978), but see Hutterer (1993).

IUCN Category of Threat: Lower Risk (subcategory Least Concern).

Distribution: This species ranges from the Amur (Russian Federation) and Ussuri (China) rivers south through Manchuria to Korea.

Sado mole
(*Mogera tokudae*)

Taxonomy: *Mogera tokudae* Kuroda 1940. Considered a subspecies of *M. robusta* by Corbet (1978), but see Hutterer (1993).

IUCN Category of Threat: Endangered (B1 and 2c).

Description: General external appearance is similar to *Mogera minor*, but the body size is much larger; head and body length is 153–182mm and body weight is 95–164g. Upper incisor row is V-shaped and markedly projected forward.

Distribution: This species is found only on Sado Island and part of Niigata (formerly Echigo Plain), in Honshu, Japan.

Habitat: The Sado mole prefers wet plains with soft, deep soil. The peripheral population of *M. tokudae* in Niigata confronts, with a sharp line of boundary demarcation, that of *M. minor*. In spite of the very large body size (twice that of *M. minor*), *M. tokudae* is ecologically inferior to *M. minor* and the peripheral population abutting that of *M. minor* in Niigata is retreating with the expansion of the range of the smaller *M. minor* (Abe, pers. comm.). Interspecific competition may therefore eliminate *M. tokudae* from Honshu, although it may remain on Sado Island, where there are no potential competitors.

Ecology and behaviour: This species feeds on earthworms, insects, centipedes, leeches and plant seeds. Litter size ranges from 2–6, with an average of three (Abe, pers. comm.).

Japanese mole
(*Mogera wogura*)

Taxonomy: *Mogera wogura* Temminck 1842. Formerly included *M. minor*; but see Hutterer (1993).

IUCN Category of Threat: Lower Risk (subcategory Least Concern).

Description: Body size varies geographically: head and body length ranges from 125–185mm; body weight from 48–175g. Forward projecting degree of the upper incisor row is small and the shape of the row is rounded arc-like.

Distribution: This species ranges from the Ussuri and Amur rivers through Manchuria and Korea to the southern half of the main islands of Japan, southern Honshu, Shikoku, Kyushu, Oki, Tsushima, Goto, Yaku and Tanegashima (Abe, pers. comm.).

Habitat: The Japanese mole prefers wet grassy plains with deep, soft soils.

Ecology and behaviour: This species is solitary and active throughout the day. It feeds on a variety of insects, earthworms, amphibians and plant seeds. Females bearing 3–5 embryos have been found from April to June. The maximum longevity in the wild is 3.5 years (Abe, pers. comm.).

GENUS *NESOSCAPTOR*

This recently described monotypic genus is of uncertain affinities within the Talpinae.

Ryukyu mole
(*Nesoscaptor uchidai*)

Taxonomy: *Nesoscaptor uchidai* Abe, Shiraishi and Arai 1991.

IUCN Category of Threat: Endangered (B1 and 2c).

Distribution: This species has only recently been described from the Ryuku islands, Senkaku islands and the west coast of Uotsuri-jima, Japan (Abe *et al.*, 1991).

Habitat: The preferred habitat of this species is not known.

Ecology and Behaviour: A newly described species, the behaviour of *N. uchidai* awaits further investigation.

GENUS *PARASCAPTOR*

This monotypic genus was formerly included in *Talpa*; see Hutterer (1993).

Parascaptor leucura

Taxonomy: *Parascaptor leucura* Blyth 1850. Previously classified as *T. micrura* by Ellerman and Morrison-Scott (1951), but see Corbet (1978) and Hutterer (1993).

IUCN Category of Threat: Lower Risk (subcategory Least Concern).

Distribution: Assam (India), Myanmar and Yunnan (China).

Habitat: Montane forests at an altitudinal range of 1000–2500m.

GENUS *SCAPANULUS*

The genus *Scapanulus* is represented by a single species, the Gansu mole (*S. oweni*) which is restricted to China.

Gansu mole
(*Scapanulus oweni*)

Taxonomy: *Scapanulus oweni* Thomas 1912.

IUCN Category of Threat: Lower Risk (subcategory Least Concern).

Description: The moles of this genus show certain structural features similar to those of North American moles of the genera *Scapanus* and *Scalopus* in the reduction of the upper canines and enlargement of the anterior incisors. The external appearance of *Scapanulus* is mole-like, but the forelimbs, although broader than in *Scaptonyx*, are not as wide proportionately as in *Talpa*. The claws of the Gansu mole are slender, but long and flattened. The snout is rather long, tapering and grooved on the underside. Body length ranges from 98–108mm and the tail length is approximately 35–38mm. The tail is relatively thick and covered in fur. The general colour is drab grey, with a silvery appearance. Individual hairs are actually slate grey with brown tips.

Distribution: The single species in this genus has only been recorded from the provinces of South Gansu, East Qinghai, south-west Shaanxi and North Sichuan (China).

Habitat: This species appears to prefer a montane habitat, having been recorded at altitudes between 2700–3000m.

Ecology and behaviour: The habits of these moles have not been recorded. Only about six specimens exist in museums.

GENUS *SCAPTOCHIRUS*

The genus *Scaptochirus* is represented by just a single species, the short-faced mole (*S. moschatus*) which is only found in China.

Short-faced mole
(*Scaptochirus moschatus*)

Taxonomy: *Scaptochirus moschatus* Milne-Edwards 1867. This species was placed in the genus *Talpa* by Corbet (1978), but see Hutterer (1993).

IUCN Category of Threat: Lower Risk (subcategory Least Concern).

Description: This is a distinctive species, characterised by reduced premolars above and below (3/3) and a very short, wide rostrum. It also has an especially soft, lustrous pelage.

Distribution: This species occurs in north-eastern China from where it ranges south to about 36°N. It has been recorded from the following provinces: Hopei, Shantung, Shansi and Shensi. Fossil evidence has been recorded from the Holocene cave site at Shenxian, Jiangxi Province, South China and from the Lower Pleistocene period in Sichuan.

Habitat: The preferred habitat of this species is unknown.

Ecology and behaviour: No details of the behaviour of this species in the wild are available at the present time.

GENUS *SCAPTONYX*

The genus *Scaptonyx* is represented in China and Myanmar by a single species – the long-tailed mole (*S. fusicaudus*). Little is known about this species.

Long-tailed mole
(*Scaptonyx fusicaudus*)

Taxonomy: *Scaptonyx fusicaudus* Milne-Edwards 1872. A single subspecies, *S.f. affinis* Thomas, has been described from north-west Yunnan, China.

IUCN Category of Threat: Lower Risk (subcategory Least Concern).

Description: The general body form is mole-like, with the forefeet only slightly broadened, but bearing stout digging claws. The ears are very much reduced in size and well concealed within the dense fur. The tail is short and thick, and thinly covered with short stiff hairs. Body length is approximately 65–90mm and tail length about one-third this length. Fur is soft and velvety in texture; the hairs are dark slate in colour, with brown tips.

Distribution: The single species in this genus has been recorded in China – Shaanxi, East Qinghai, Sichuan, Yunnan – and northern Myanmar.

Habitat: Montane habitat at altitudes of 2100–4100m.

Ecology and behaviour: Nothing is known about the ecology of this species.

GENUS *TALPA*

In the Old World, the genus *Talpa* consists of nine species which are distributed throughout temperate Eurasia. *Talpa* is the most widespread genus within the Family Talpidae, with representative species ranging from western Europe through Asia and south into the Indomalayan Region.

All members of this genus are adapted for a fossorial lifestyle; external bodily appendages are severely reduced and the body shape is almost cylindrical, tapering in a pointed snout. A short, usually sparsely furred tail is present. Forelimbs are highly developed, modified to loosen and excavate earth in the process of tunnel formation. All species create a complex, three-dimensional layer of interconnecting tunnels. These provide shelter from the elements, reduce predation but, most importantly, act as a food trap for earthworms and insect larvae which migrate through the soil column.

Siberian mole (*Talpa altaica*)

Taxonomy: *Talpa altaica* Nikolsky 1883.

IUCN Category of Threat: Lower Risk (subcategory Least Concern).

Description: Elongate body with uniformly short, usually black, fur. Prominent features are the broad, spade-like forelimbs, pink fleshy snout and short tail. Ear pinnae are absent. The tapering head is set deeply into the main body giving the appearance of the absence of a proper neck. Seldom seen on the surface and unlikely to be confused with other species. Body length of *T. altaica* 136–203mm. The fur in this species is relatively long with a brownish hue on the back.

Distribution: The taiga zone of Siberia between the rivers Ob/Irtysh and Lena, reaching north along the Yenesei river to 70°N and south to northern Mongolia.

Mediterranean/Blind mole (*Talpa caeca*)

Taxonomy: *Talpa caeca* Savi 1822.

IUCN Category of Threat: Lower Risk (subcategory Least Concern).

Distribution: Parts of Iberia, southern side of the Alps, Balkans, Thracia (Greece); perhaps Asia Minor.

Ecology and behaviour: Unknown, probably similar to *T. europaea*. This species is marginally sympatric with *T. europaea* in the Alps.

Caucasian mole (*Talpa caucasica*)

Taxonomy: *Talpa caucasica* Satunin 1908. Included in *T. europaea* by Ellerman and Morrison-Scott (1951) and by Kuzyakin (1965), but considered a distinct species by Gromov *et al.*, (1963) and by Kozlovsky *et al.*, (1972), the latter authors basing their view on karyological data (Corbet, 1978).

IUCN Category of Threat: Lower Risk (subcategory Least Concern).

Description: As for *T. altaica*. Fur is short and of a uniform length. Eyes completely covered with delicate thin skin.

Distribution: Northern side of the Caucasus, from the Sea of Azov to the Caspian Sea.

European mole (*Talpa europaea*)

Taxonomy: *Talpa europaea* Linnaeus 1758.

IUCN Category of Threat: Lower Risk (subcategory Least Concern).

Description: Generally as for *T. altaica*. Fur is short and of a uniform length. See Godfrey and Crowcroft (1978) for general description.

The European mole (*Talpa europaea*) is one of the few insectivores to have been hunted for its fur. Changing agricultural practices are a threat to several species of fossorial insectivores. (Photo by David Stone)

Distribution: Widespread throughout Europe except for Ireland, Scandinavia (except southern Sweden) and parts of Mediterranean coastline. In the east it ranges throughout Russia to as far as the rivers Ob and Irtish.

Habitat: The European mole is an adaptable species and is present in most habitats where the soil is sufficiently deep to allow tunnel construction. Originally inhabitants of deciduous woodlands (in western Europe), moles have taken advantage of agricultural developments and thrive in pasture and arable lands which support high densities of soil invertebrates, especially earthworms. May occur in coniferous forests, but usually have much larger ranges. Rarely found on moorland, sand dunes and similar poor soils. Found up to 2000m in the Alps.

Ecology and behaviour: The European mole is specialised to a fossorial life, constructing a permanent series of tunnels in which the animal may spend its entire lifetime. Active day and night; in some cases with a cycle of 3–4 hours activity and 3–4 hours rest. Structured system of social and spatial organization under certain conditions. This species is widely persecuted in parts of its range – where it interferes with agriculture, particularly cereals and market gardening activities (see Stone, 1989). The impacts of such events are restricted to a local or regional scale. This species has been relatively well studied in the wild; see Godfrey and Crowcroft (1960), Stone (1986) and Gorman and Stone (1990) for reviews.

Levant mole
(*Talpa levantis*)

Taxonomy: *Talpa levantis* Thomas 1906. Formerly considered a subspecies of *T. caeca* by Corbet (1978); but see Hutterer (1993).

IUCN Category of Threat: Lower Risk (subcategory Least Concern).

Distribution: Bulgaria, Thracia (Greece), North Anatolia (Turkey) and adjacent Caucasus.

Iberian mole
(*Talpa occidentalis*)

Taxonomy: *Talpa occidentalis* Thomas 1902. Formerly considered a subspecies of *T. caeca* by Corbet (1978); but see Hutterer (1993).

IUCN Category of Threat: Lower Risk (subcategory Least Concern).

Distribution: West and Central Iberian Peninsula.

Roman mole
(*Talpa romana*)

Taxonomy: *Talpa romana* Thomas 1902. Formerly included *T. stankovici*; but see Hutterer (1993).

IUCN Category of Threat: Lower Risk (subcategory Least Concern).

Description: Similar to *T. europaea*.

Distribution: Appenine region of Italy, and extreme south-eastern France; formerly Sicily.

Habitat: The preferred habitat of this species is grassland and woodland.

Ecology and behaviour: From preliminary field data the social behaviour of this species appears somewhat similar to that of *T. europaea* (Loy, Dupré and Stone, 1992).

Balkan mole
(*Talpa stankovici*)

Taxonomy: *Talpa stankovici* V. and E. Martino 1931. Formerly considered a subspecies of *T. romana* by Corbet (1978), but see Hutterer (1993).

IUCN Category of Threat: Lower Risk (subcategory Least Concern).

Distribution: European Balkans, Greece (including Corfu Island), Macedonia and probably Albania.

Persian mole
(*Talpa streeti*)

Taxonomy: *Talpa streeti* Lay 1965

IUCN Category of Threat: Critically Endangered (B1 and 2c)

Description: Very similar in appearance to *T. romana* but the teeth appear distinctive. Only the type specimen exists.

Distribution: Known only from the type locality, Hezer Darrak in Kurdistan Province, north-west Iran.

GENUS *UROTRICHUS*

Two species are recognised in this genus: *U. pilirostris* and *U. talpoides*. Both are restricted to Japan where they occur in montane habitats, particularly regions of coniferous forest up to 2000m. Both species are reported to be plentiful but do not occur on the plains.

Lesser Japanese shrew-mole (*Urotrichus pilirostris*)

Taxonomy: *Urotrichus pilirostris* True 1886.

IUCN Category of Threat: Lower Risk (subcategory Least Concern).

Description: Shrew-like in appearance: the forelimbs are only slightly broadened, ears are present but are small and concealed within the fur. The tail is also densely covered in coarse hairs and is often enlarged with fat. The fur is soft and dense but coarser in texture than other talpids. Fur colour varies from dark brown to black, with a metallic sheen in reflected light. Body length measures 40–84mm with a tail length of 32–44mm (Abe, pers. comm.). This species burrows extensively in soil as well as under leaf litter.

Distribution: This montane species is found on the islands of Honshu, Shikoku and Kyushu, Japan.

Habitat: Montane coniferous forest at high altitudes. When two species of shrew-moles occur on the same mountainside, *U. pilirostris* is always found at higher altitudes (Abe, pers. comm.).

Ecology and behaviour: The diet of this species consists of insects, spiders, worms and other invertebrates. Shrew-moles breed in April and May and the average litter size is three. The gestation period is not known. Little else is known about the ecology of this species.

Greater Japanese shrew-mole (*Urotrichus talpoides*)

Taxonomy: *Urotrichus talpoides* Temminck 1841.

IUCN Category of Threat: Lower Risk (subcategory Least Concern).

Description: Similar in external appearance to *U. pilirostris*, but slightly larger body with a shorter tail.

Distribution: This species is restricted to the Japanese islands of Honshu, Shikoku, Kyushu, Dogo and North Tsushima.

Habitat: Forest and grasslands at low altitude.

Ecology and behaviour: The diet mainly consists of insects, earthworms and centipedes. Maximum life span appears to be 3.5 years. Greater shrew-moles breed from March to May. The litter size ranges from 2–4, with an average of 3.1 (Abe, pers. comm.).

This species burrows just beneath the surface but *U. talpoides* has also been recorded foraging on the surface and even observed to climb low bushes. In winter this species is often found dead in bird nest boxes in trees at heights of 2–4m above the ground. It is likely, however, that these specimens have been carried there by avian or small terrestrial predators.

Sub-family Uropsilinae

GENUS *UROPSILUS*

The genus *Uropsilus* was first described by Milne-Edwards (1871) from material sent back from Muping (now Baoxing), Sichuan, by the renowned French missionary, Père Armand David. These small insectivores are shrew-like in appearance, but exhibit a mole-like skull and dentition. Four species are now recognised in this genus (sub-family Uropsilinae) (Hutterer, 1993). Previously only one species was recognised (Ellerman and Morrison-Scott, 1951) on the grounds that the variability of dental formula was erratic. But in a detailed revision of the genus, Hoffmann (1984) demonstrated that three distinct species, with areas of sympatry, could be recognised. Wang and Yang (1989) have since recognised a fourth species. All members of this genus are restricted to southern China and adjacent parts of Myanmar. Almost nothing is known about their ecology.

Uropsilus andersoni

Taxonomy: *Uropsilus andersoni* Thomas 1911. Two forms described as subspecies of *U. andersoni* are conspecific with *U. gracilis* (Hoffmann, 1984).

IUCN Category of Threat: Lower Risk (subcategory Least Concern).

Description: Similar in external appearance to *U. soricipes*.

Distribution: This species is restricted to South (Corbet, 1992) and Central (Hutterer, 1993) Sichuan, China.

Habitat: Forest and alpine regions.

Ecology and behaviour: The conservation requirements and behaviour of this species in the wild are poorly known.

Uropsilus gracilis

Taxonomy: *Uropsilus soricipes* Thomas 1911

IUCN Category of Threat: Lower Risk (subcategory Least Concern).

Description: Similar in external appearance to *U. soricipes*, but with a longer tail, Considerable geographical variation has been noted in the size of *U. gracilis* (Hoffmann, 1984).

Distribution: South Shaanxi, Sichuan and western Yunnan (China) and North Myanmar at altitudes of 1200–4250m. Sympatric with *U. soricipes* in parts of Central Sichuan.

Habitat: Forest and alpine regions.

Ecology and behaviour: The behaviour and conservation requirements of this species are still poorly known.

Uropsilus investigator

Taxonomy: *Uropsilus investigator* Thomas 1922. Formerly included in *U. gracilis* by Hoffmann (1984). On the basis of morphology and distribution, Wang and Yang (1989) concluded that both are sympatric in Yunnan and should therefore be regarded as distinct species.

IUCN Category of Threat: Endangered (B1 and 2c).

Description: This species measures from 67–83mm, with a tail length of 54–75mm.

Distribution: This species has only been recorded from Yunnan, People's Republic of China, specifically the Kui-chiang-Salween divide at 28°N (Wang and Yang, 1989). It is known only from the type specimen, collected at 3600m.

Habitat: The habitat preferences of this species are not known. They are possibly similar to those of *U. soricipes*.

Ecology and behaviour: The ecology of this species has not been investigated in the wild.

Chinese shrew-mole (*Uropsilus soricipes*)

Taxonomy: *Uropsilus soricipes* Milne-Edwards 1872.

IUCN Category of Threat: Endangered (B1 and 2c).

Description: In appearance this species measures from 63–88mm, with a long tail of 54–75mm. Colour is apparently quite variable, ranging from dark brown to slate-grey. Overall, this species resembles a shrew more than a mole and its unspecialised limbs suggest that it probably forages beneath leaf litter rather than constructing a network of tunnels. The long, scaly snout is formed of two tubular nostrils with a groove along the top. The ears extend beyond the fur on the head and are conspicuous. The tail is slender and covered with rings of small scales.

Distribution: This species is restricted to a small area of Central Sichuan at altitudes of 1500–2700m (Hoffmann, 1984).

Habitat: Forest and alpine regions. Sympatric with *U. gracilis* in some areas.

Ecology and behaviour: The behaviour of this species in the wild still awaits detailed examination.

Chapter 3

The Scandentia of Asia

3.1 Introduction

The tree shrews (Order Scandentia, Family Tupaiidae) comprise a small number of generally well-defined species that are only found in South and south-east Asia. Five genera (19 species) are currently recognised (Wilson, 1993) (see Table 3.1). All occur in forested areas, ranging from India and south-west China eastward through Malaysia, Indonesia (west of Wallace's Line) and the Philippines.

In appearance, tree shrews resemble long-snouted squirrels (the Malay word *"Tupai"* means squirrel) but, with the exception of the pen-tailed tree shrew (*Ptilocercus lowi*), tree shrews can be readily distinguished from squirrels by the absence of long facial whiskers. *Ptilocercus* itself is easily identified by its tail, which is naked except for a whitish feather-shaped arrangement of the hairs near the end.

All tree shrews are of a slender build, adults generally weighing 70–100g. The length of the head and body ranges from 100–220mm, while tail length varies from 90–225mm. Generally a russet-brown colour, they have a long, pointed muzzle with 38 sharp, pointed teeth. The pelage is of the normal mammalian type, consisting of long, straight guard hairs and shorter, softer, more woolly underfur. Some forms have pale shoulder stripes and others have facial markings. The ears are squirrel-like; that is, they are comparatively small and cartilaginous, except in the pen tailed tree shrew in which they are larger and more membranous. The feet of tree shrews are naked beneath; the soles are adorned with tubercle-like pads which assist with climbing. The long and supple digits bear sharp, moderately curved claws.

In addition to their external resemblance to squirrels, tree shrews are mostly diurnal and some of their actions and movements are similar. *Ptilocercus* differs in being mainly nocturnal and, when on the ground, progresses in a series of hops. Other tree shrews are swift runners. All tupaiids are capable climbers, seeking their food in trees as well as on the ground. Their diet consists mainly of insects and fruit, but occasionally includes other animal food and various types of plant matter. They are generally fond of water for both drinking and bathing.

Surprisingly, in view of their large size (relative to insectivores) and diurnal habits, tree shrews have attracted relatively little biological attention. As a result there has been little work on their ecology or behaviour in their natural environments.

The reproductive behaviour of most species is still poorly known. *Tupaia* spp. apparently breed throughout the year. Female *Urogale everetti* are receptive to males soon after giving birth. The known gestation periods are 46 to 50 days for *T. glis* and probably 54–56 days for *U. everetti*. Female tree shrews have one, two or three pairs of mammae, and litter size varies from 1–4.

Tree shrews have well developed senses of vision, hearing and smell. Some of the primate-like features of tree shrews are the relatively large brain case, the similarity of the carotid and subclavian arteries to those of man, and the permanent scrotal sac in males. The orbits are also encircled by bone, as in other primates and certain other mammals. The upper incisors are large, canine-like and separated and the canines are small and resemble the premolars. The upper molars are broad with a W-shaped patterns of cusps. The dental formula is: i2/3, c1/1, pm3/3, m3/3 × 2 = 38.

Most species of tree shrew appear to be tolerant of some degree of habitat disturbance, although the extent of this tolerance has not yet been measured in the field. The conservation status of most species is also unclear. The status (both taxonomic and ecological) of subspecies is even less well known. Several forms occupy restricted areas, many of which are undergoing transformation as a result of human interference. Until recently it was generally viewed that most tree shrews were not in any immediate threat of extinction; none have been listed in the IUCN Red Lists to date. However, following the application of recent IUCN criteria to known distributions (see Box 2.1 and Appendix III), several species are now recommended as "Vulnerable" and two species, *Tupaia nicobarica* and *T. longipes* have even been classified as "Endangered" (see below). All of these species occupy restricted ranges, often islands, where agricultural encroachment and/or logging is a major activity and increasing threat as a result of habitat loss and alteration.

Habitat conservation is vital to preserving tree shrews. The forests that are important to tree shrews are also known to be of great importance to a wide range of plants, mammals, birds and invertebrates, many of which are endemic. By developing and implementing conservation strategies which include tree shrew conservation requirements in these forests, the overall biological diversity of these unique ecosystems could be far better protected.

3.2 Taxonomic Classification

Few other mammalian families have proved as difficult to classify as the tree shrews. Historically, the Tupaiidae have been grouped with both the Macroscelididae (elephant shrews) and the insectivore suborder Menotyphla. Simpson (1945) recognised the Tupaiidae as belonging to the superfamily Tupaioidea, within the infraorder Lemuriformes of the primate suborder Prosimii. Recently, however, there has been an increasing tendency to omit the tree shrews from classifications of the primates (Hershkovitz, 1977; Petter and Petter-Rousseaux, 1979). Studies of behaviour and reproduction led Martin (1968) to consider the Tupaiidae as advanced insectivores intermediate to Lipotyphlans and Primates (Walker, 1991). More recent studies have suggested that the Tupaiidae have no immediate relationship with the Macroscelididae and should be placed in a separate order closely related to the Primates, Chiroptera and Dermoptera (Dene, Goodmann and Prychodko, 1978; Goodman, 1975; Luckett, 1980; McKenna, 1975). Such an order, with the name Scandentia, has now received general recognition (Corbet and Hill, 1986; Honacki *et al.*, 1982; Napier and Napier, 1985; Yates, 1984).

All members of the Scandentia are restricted to southeast and South Asia. Nineteen species are generally recognised within this order, comprising five genera (Table 3.1): *Anathana*, *Dendrogale*, *Tupaia*, *Urogale* and *Ptilocercus*. Three genera and 10 species occur on the island of Borneo alone. Of the remaining two genera, *Anathana* (a single species, *A. elliotti*) occurs only in southern and central India, while *Urogale* (one species, *U. everetti*) occurs only on Mindanao, Dinagat and Siargao islands in the southern Philippines.

3.3 Species Accounts

Sub-family Tupaiinae

GENUS *ANATHANA*

This genus is represented by a single species in India. Very little is known about the ecology, distribution or conservation status of this species.

Madras tree shrew (*Anathana elliotti*)

Taxonomy: *Anathana elliotti* Waterhouse 1850.

IUCN Category of Threat: Lower Risk (subcategory Least Concern).

Description: This species differs from *Tupaia* in its larger and thicker-haired ears, as well as in dental and skull features. In size, the body measures from 175–200mm, with a tail length of 160–190mm. Body weight is about 160g (Martin, 1984). In colour, the upper parts are usually speckled yellow and brown, with the middle of the back, the rump and sometimes the upper tail tinged with reddish. The underside is usually a whitish colour. A cream-coloured shoulder stripe is present.

Table 3.1. Classification of tree shrews (Wilson, 1993)

Genus	Species
Sub-family Tupaiinae	
Anathana	A. elliotti
Dendrogale	D. melanura
	D. murina
Tupaia	T. belangeri
	T. chrysogaster
	T. dorsalis
	T. glis
	T. gracilis
	T. javanica
	T. longipes
	T. minor
	T. montana
	T. nicobarica
	T. palawanensis
	T. picta
	T. splendidula
	T. tana
Urogale	U. everetti
Sub-family Ptilocerinae	
Ptilocercus	P. lowi

Distribution: This species is only found in the forested regions of Peninsular India, north as far as Bihar, in the east and the Satpura Hills (south-west Madhya Pradesh) in the west.

Habitat: Occurs to at least 1400m (Chorazyna and Kurup, 1975).

Ecology and behaviour: The habits of this species are not well known but are probably similar to those of *Tupaia* and *Urogale*. Unlike *Tupaia*, however, this species does not frequent human dwellings (Roonwal and Mohnot, 1977).

GENUS *DENDROGALE*

There are two species of small smooth-tailed tree shrew: *Dendrogale melanura* and *D. murina*, which are geographically isolated (Lekagul and McNeely, 1977; Medway, 1977). These are the only small members of the family Tupaiidae with round, uniformly even-haired tails. Shoulder stripes are not present. Both species are active during the day.

Bornean smooth-tailed tree shrew (*Dendrogale melanura*)

Taxonomy: *Dendrogale melanura* Thomas 1892. Two subspecies are recognised: *D.m. melanura* and *D.m. baluensis* (Payne, pers. comm.).

IUCN Category of Threat: Vulnerable (B1 and 2c).

Description: Body length is 110–150mm, tail length 90–140mm. Upper parts are a mixed blackish and ochraceous-buff or cinnamon-rufous colour, while under parts and the inner side of the legs are ochraceous. There are no distinct facial markings as in *D. murina*. The claws of this species are long. *D.m. melanura* is reportedly darker above and more reddish below than *D.m. baluensis*.

Distribution: This species is restricted to the montane regions of Borneo, where it occurs at an elevation of 900–1500m. *D.m. melanura* has been recorded from Mt Dulit, Mt Mulu and the Kelabit highlands in northern Sarawak as well as from the Sarawak-Sabah border. *D.m. baluensis* has been recorded from Mt Kinabalu and Mt Trus Madi in Sabah.

Habitat: Montane forest.

Ecology and behaviour: Little is known about the behaviour of this species in the wild. It is diurnal.

Northern smooth-tailed tree shrew (*Dendrogale murina*)

Taxonomy: *Dendrogale murina* Schlegel and Müller 1843.

IUCN Category of Threat: Lower Risk (subcategory Least Concern).

Description: Similar in size to *D. melanura*. In appearance, this species is light in colour with distinctive facial markings: the side of the head has a blackish line which extends from the base of the nasal vibrissae through the eye to the ear, with light, buff coloured lines above and below the black band. Upper parts vary in colour from brown to black, with a buff coloured underside. The claws are small. (See Lekagul and McNeely (1977) for detailed description.)

Distribution: This species is restricted to southern Vietnam, Cambodia and eastern Thailand. It may be quite common in the forests along the Cambodian border, but this shy species is often inconspicuous.

Habitat: Montane forest.

Ecology and behaviour: Little is known abut the ecology of *D. murina*. It is a diurnal species living in the lower forest canopy, but also foraging on the ground. These animals are more arboreal than *T. glis*, with which they are sympatric (Lekagul and McNeely, 1977).

GENUS *TUPAIA*

The genus *Tupaia* is the most well-represented and widespread genus of tree shrews, with 14 species currently recognised.

Northern tree shrew (*Tupaia belangeri*)

Taxonomy: *Tupaia belangeri* Wagner 1841. Variation is considerable within this species although there has been no comprehensive review of its members. *T. belangeri* was formerly included in *T. glis* by Ellerman and Morrison-Scott (1951) but more recently has been given specific rank (see Corbet, 1992).

IUCN Category of Threat: Lower Risk (subcategory Least Concern).

Distribution: This species occurs in Burma and Thailand, reaching as far south as the Isthmus of Kra. Its range also extends north through Indochina to Guangxi, Yunnan and southern Sichuan Provinces in China, and northwest to Assam (India), Bangladesh and East Nepal. It also occurs on Hainan Island (China). *Tupaia belangeri* has been recorded from an altitude of 2300m in Burma and 2500m in Yunnan Province (China).

Golden-bellied tree shrew (*Tupaia chrysogaster*)

Taxonomy: *Tupaia chrysogaster* Miller 1903. Considered a subspecies of *T. glis* by Corbet (1992); but see Wilson (1993). A member of the endemic Mentawi Islands fauna; may also include *siberu* and *tephrura* from Siberut and Tana Bala Islands (Wilson, 1993).

IUCN Category of Threat: Vulnerable (B1 and 2c).

Distribution: Tana Bala, Siberut, Sipora, North and South Pagi to the west of Sumatra (Indonesia).

Ecology and behaviour: Few details are available on this localised species.

Striped tree shrew (*Tupaia dorsalis*)

Taxonomy: *Tupaia dorsalis* Schlegel 1857. Often placed in the same genus as *Lyonogale* (= *Tana*) (Corbet, 1992).

IUCN Category of Threat: Lower Risk (subcategory Least Concern).

Description: The striped tree shrew is speckled olive-brown in appearance. A thin black line extends from the neck to the base of the tail. A white shoulder strip is usually a distinctive feature. Ventral coloration is grey-buff (Payne *et al.*, 1985).

Distribution: This species appears to be widespread throughout the lowlands of Borneo up to 1000m, but usually occurs at lower altitudes. There are no records from the south-east part of the island (Payne, pers. comm.).

Habitat: Lowland tall and secondary forest.

Ecology and behaviour: Diurnal, mainly terrestrial, in primary and secondary forest (Payne *et al.*, 1985). A rarely seen species.

Common tree shrew (*Tupaia glis*)

Taxonomy: *Tupaia glis* Diard 1820. Many subspecies and races have been described on the basis of colour and size (see Corbet, 1992). Most forms have been reviewed by Hill (1960). In Malaysia, dorsal coloration becomes paler towards the north, both on the mainland and islands. Many subspecies are listed across the range of this species (see Wilson, 1993).

IUCN Category of Threat: Lower Risk (subcategory Least Concern).

Description: Hairs on the upper part are banded dark and pale appearing finely speckled brown or reddish brown. Usually has a pale stripe on each shoulder (Payne *et al.*, 1985).

Distribution: This species occupies a very wide distribution: from south of the Isthmus of Kra (Thailand) through the Malay Peninsula to Indonesia (Sumatra and Java), and on the following small islands:

- Samui, Pannan, Perhentian, Refang, Tioman, Pemangil and Aor on the east coast of Malaysia;

- Panjang, Telibon, Lankawi, Terutau and Penang on the west coast of Malaysia; and

- Singapore, Batam, Bintang, Mapor, Lingga, Sinkep and Banka on the south coast of Malaysia.

Habitat: *T. glis* occurs widely in plantations and gardens, as well as natural forest. It has been recorded up to 1120m in Borneo and 1420m on Peninsular Malaysia.

Ecology and behaviour: This species is more terrestrial than arboreal, feeding mainly on invertebrates but also fruit. It is a diurnal, territorial species (Langham, 1982; Payne *et al.*, 1985). The natural population density of *T. glis* has been reported as 6–12/ha in Thailand and 2–5/ha in Peninsular Malaysia (Langham, 1982; Lekagul and McNeely, 1977; Sorenson, 1970, 1974). The pair bond between a male and female seems pronounced (Martin, 1968). In a study of *T. glis* in Singapore, Kawamichi and Kawamichi (1979, 1982) found males to pair with 1–3 females. In this area there was little overlap between the home ranges of adult residents of the same sex, but those of opposite sexes overlapped completely and a male's range sometimes included the ranges of more than one female. Home range size in the study area averaged 10,174m^2 for males and 8809m^2 for females.

Slender tree shrew (*Tupaia gracilis*)

Taxonomy: *Tupaia gracilis* Thomas 1893. Three full subspecies have been recognised: *T. g. edarata* from Telok

Edar and Karimata Island (smaller, hind foot <40mm, dorsal pelage brownish); *T.g. gracilis* from mainland Borneo and Banggi (large, hind foot >40mm, less russet above) and *T.g. inflata* from Banka and Billiton (smaller, hind foot <40mm, more russet on posterior back).

IUCN Category of Threat: Lower Risk (subcategory Least Concern).

Description: This species resembles *T. minor* in appearance: dorsal hairs are speckled olive-brown and the ventral coloration is usually buff-white with no reddish tinge.

Distribution: *T. gracilis* is confined to Borneo and adjacent small islands. Recorded from the lowlands and hills in most areas except the south-east between Riam in Central Kalimantan, and South Mahakam in East Kalimantan. It also occurs on the islands of Banggi, Karimata, Banka and Billiton.

Habitat: Tall and secondary forests. Occurs up to 1200m in the Kelabit highlands, but usually prefers lower altitudes (Payne *et al.*, 1985).

Ecology and behaviour: Diurnal and arboreal (Payne *et al.*, 1985). No other details of its ecology are known.

Javan tree shrew
(*Tupaia javanica*)

Taxonomy: *Tupaia javanica* Horsfield 1822. Several subspecies have been proposed on the basis of colour variations, but this is not judged to be a reliable distinguishing character, even within the same locality (Hill, 1960).

IUCN Category of Threat: Lower Risk (subcategory Least Concern).

Distribution: Bali, Java, West Sumatra and the Island of Nias (Indonesia).

Habitat: The habitat requirements of this species are not known.

Bornean tree shrew
(*Tupaia longipes*)

Taxonomy: *Tupaia longipes* Thomas 1893. Considered a subspecies of *T. glis* by Corbet (1992); but see Wilson (1993).

IUCN Category of Threat: Endangered (B1 and 2c).

Distribution: Siantan, Riabu and Jimaja islands off the east coast of Peninsular Malaysia.

Pygmy/Lesser tree shrew
(*Tupaia minor*)

Taxonomy: *Tupaia minor* Gunther 1876. Five subspecies are recognised: *T.m. caedis* (North Borneo); *T.m. humeralis* (West Sumatra); *T.m. malccana* (Lingga Island); *T.m. minor* (Borneo) and *T.m. sincipis* (Sinkep Island).

IUCN Category of Threat: Lower Risk (subcategory Least Concern).

Description: As its name suggests, this species is one of the smallest tree shrews (weighing 30–70g). Its small size and long thin tail are usually distinctive features. Dorsal hairs are banded light and dark, giving an overall speckled olive-brown appearance. Ventral coloration is usually a buff colour, often with a reddish tinge towards the rear (Payne *et al.*, 1985). *T.m. minor* is reported to differ from *T.m. caedis* in having a wider, whiter shoulder stripe and browner underparts.

Distribution: This species is found in Peninsular Thailand, Peninsular Malaysia, Sumatra, Borneo and the smaller islands of Singkep Lingga, Banggi, Balembangan and Laut. On Borneo, *T.m. minor* is found throughout the lowlands and hills except in the north-east. *T.m. caedis* has been recorded from eastern Sabah from Kudat to Kalabakan, as well as Banggi and Balambangan islands (Payne *et al.*, 1985).

Habitat: In Borneo, found in plantations, forest and gardens up to 1700m.

Ecology and behaviour: Diurnal and arboreal. Feeds on insects and fruit (D'Souza, 1974; Payne *et al.*, 1985).

Mountain tree shrew
(*Tupaia montana*)

Taxonomy: *Tupaia montana* Thomas 1892. Two subspecies have been described: *T.m. baluensis* and *T.m. montana*.

IUCN Category of Threat: Lower Risk (subcategory Least Concern).

Description: Considerable colour variation exists in *T. montana*, from dark brown to reddish- or olive-brown, but always with a fine reddish speckling. Underfur is usually buff-red on grey. Some individuals exhibit a pale shoulder stripe. *T.m. montana* has a black stripe on the underparts which narrows between the shoulders and broadens between the hind limbs. The absence of this black patch, as well as its smaller size, distinguishes *T.m. baluensis*.

Distribution: The mountain tree shrew only occurs in north-west Borneo. *T.m. baluensis* has been recorded from Mt Kinabalu (900–3170m), Mt Trus Madi (1530–2380m) and the Crocker Range (above 1200m) in Sabah; from mountains along the Sabah-Sarawak border (above 1070m) and from Mt Murud (up to 2140m), the Kelabit highlands (above 1130m), Mt Mulu (above 1220m), upper S. Rajang (above 900m), Mt Penrisen (1160m), Mt Pueh and Mt Sidong (370m) in Sarawak. *T.m. montana* has been recorded from the more isolated mountain areas of northern Sarawak, including Batu Song (900m), Mt Kalulung, Mt Dulit (above 600m) and Usun Apau (Payne *et al.*, 1985).

Habitat: Montane forests at altitudes from 370–3170m.

Ecology and behaviour: Diurnal and mainly terrestrial: travels and feeds on the ground. Diet comprises a mixture of plant and animal materials (Payne *et al.*, 1985). One of the commonest mammals in primary montane forests in Sabah.

Nicobar tree shrew
(*Tupaia nicobarica*)

Taxonomy: *Tupaia nicobarica* Zelebor 1869. A subspecies, *T.n. surda*, has been described from Little Nicobar Island on the basis of a duller, less yellow colour.

IUCN Category of Threat: Endangered (B1 and 2c).

Distribution: This species occurs on both Great and Little Nicobar Islands in the Indian Ocean.

Habitat: The preferred habitat of this species is not known.

Ecology and behaviour: No additional information available at the present time.

Palawan tree shrew
(*Tupaia palawanensis*)

Taxonomy: *Tupaia palawanensis* Thomas 1894. Provisionally considered a subspecies of *T. glis* by Corbet (1992); but see Wilson (1993).

IUCN Category of Threat: Vulnerable (B1 and 2c).

Distribution: Palawan, Balabac, Culion, Busuanga (Calamian Island) and Cuyo in the southern Philippines.

Ecology and behaviour: Primarily a solitary and ground-dwelling species with peak activities occurring twice a day (Dans, 1993). Population density ranges from 1.6–3.2 individuals per hectare. Omnivorous feeders with a strong preference for insects with hard exoskeletons and fleshy fruits as supplemental food items (Dans, op. cit.).

Painted tree shrew
(*Tupaia picta*)

Taxonomy: *Tupaia picta* Thomas 1892. Two subspecies have been recorded (Payne, pers. comm.): *T.p. picta* in north-western Borneo, is brightly coloured, while *T.p. fuscior* Medway 1977, from coastal East Kalimantan is duller and smaller.

IUCN Category of Threat: Lower Risk (subcategory Least Concern).

Description: Underparts generally brown with heavy buff flecking and a black central stripe on the front half of the body. Pale buff shoulder stripe usually present. Ventral coloration usually a dull orange with pale grey bases to the fur. Underside and distal end of tail orange or reddish tinge (Payne *et al.*, 1985).

Distribution: This species occurs in the north-west and eastern regions of Borneo. All records of this species have been made below 1000m.

Ecology and behaviour: Diurnal. Otherwise unknown.

Ruddy tree shrew
(*Tupaia splendidula*)

Taxonomy: *Tupaia splendidula* Gray 1865. Two subspecies were provisionally recognised by Corbet (1992): *T.s. splendidula* from mainland Borneo and the Natuna Islands and *T.s. carimatae* from Karimata Island (North Natuna Islands).

IUCN Category of Threat: Lower Risk (subcategory Least Concern).

Description: A plain reddish tree shrew with pale orange shoulder stripe (if present). Upper parts tend to be darker in the midline and lighter on the sides; undersides dark red with an orange throat. Hairs on tail dark red above, orange below. *T.s. splendidula* coloration is dull, but *T.s. carimatae* is brighter and has a proportionately shorter tail and hindfeet.

Distribution: This species occurs in the lowlands of Borneo and adjacent islands of Bunguran, Laut (North Natuna Islands) and Karimata Island.

Habitat: Lowland forest.

Ecology and behaviour: Diurnal. Otherwise unknown.

Large tree shrew
(*Tupaia tana*)

Taxonomy: *Tupaia tana* Raffles 1821. Since the critical review of nomenclature by Lyon (1913), the following subspecies have been proposed for Borneo (Medway, 1977): *T.t. besara, T.t. speciosa, T.t. utara, T.t. nitida, T.t. kelabit, T.t. chrysura, T.t. paitana* and *T.t. kretami.*

IUCN Category of Threat: Lower Risk (subcategory Least Concern).

Description: All subspecies have a basically similar coloration: upper parts paler towards the front of the body and darker towards the rear. Dark midline also obvious. Ventral coloration reddish-buff.

Distribution: This species occurs on Sumatra and Borneo (up to 1500m) including the following islands: Tana Balu and Tana Mara (Batu Islands); Tuangku (Banjak Island); Lingga, Banggi; Sirhassen (South Natuna Island), Big Tambelan and Bunoa (Tambelan Island).

Habitat: Rarely found outside of tall forest or dense, shaded areas in secondary forest.

Ecology and behaviour: Diurnal, mainly terrestrial in dense forest. Feeds on arthropods, earthworms and fruit (Payne *et al.*, 1985).

GENUS *UROGALE*

Philippine tree shrew
(*Urogale everetti*)

This genus comprises a single species and is only found in the Philippines.

Taxonomy: *Tupaia everetti* Thomas 1892.

IUCN Category of Threat: Vulnerable (B1 and 2c).

Description: *U. everetti* is readily distinguished from other tree shrews by its elongated snout and even-haired, rounded tail, which is more densely furred than members of the genus *Dendrogale*. Philippine tree shrews measure from 170–220mm, with a tail length of 115–175mm. A fully grown male may weigh as much as 350g. In appearance the fur is a brownish colour, while the under parts are orange-red, being most striking on the chest. There is often an indistinct orange-coloured shoulder stripe.

Distribution: This species is confined to the Island of Mindanao (collected from Mt Apo), in the Philippines, as well as adjacent smaller islands of Dinagat and Siargao (Heaney and Rabor, 1982).

Habitat: Montane forest.

Ecology and behaviour: The Philippine tree shrew is omnivorous, feeding on a wide range of insects, fruit, lizards and small mammals. This species has been maintained and bred in several zoos. The gestation period is probably 54–56 days, with just 1–2 young produced. Following birth, the female is once again receptive to males.

Sub-family Ptilocercinae

GENUS *PTILOCERCUS*

The pen-tailed or feather-tailed tree shrew is the only species within the genus *Ptilocercus*.

Pen-tailed tree shrew
(*Ptilocercus lowi*)

Taxonomy: *Ptilocercus lowi* Gray 1848. A single subspecies, *P.l. continensis* Thomas 1910 has been described from near Kuala Lumpur, Peninsular Malaysia, on the basis of greyer pelage, a longer, dark eye stripe and narrower muzzle. However, too few specimens are currently available to assess the validity of these distinctions (Corbet, 1992).

IUCN Category of Threat: Lower Risk (subcategory Least Concern).

Description: In appearance the head is moderately tapering and the whiskers are elongated and rather rigid. The limbs are nearly equal in length and the hands, feet and pads are relatively larger than in the other genera of tree shrews. The five fingers and five toes bear short, sharp claws.

The head and body measure 100–140mm and the tail length ranges from 130–190mm. The fur is soft in texture, usually dark greyish brown above and yellowish-grey beneath. The tail is naked and dark except for the terminal part which has whitish hairs on opposite sides, which produces a feather-like form as all of the hairs are in the same plane. The entire tail resembles an old-fashioned quill pen, hence the common name for this species.

Distribution: This elegant looking species is found in forested regions of Peninsular Thailand, Peninsular Malaysia, Sumatra, North and north-west Borneo and small adjacent islands of Labuan, Sirhassen, Banka, Pinie (Batu Islands), Kariman and Siberut (Mentawai Islands).

Habitat: Largely canopy-dwelling animals, pen-tailed tree shrews are found in forests and gardens. They are much less at ease when on the ground compared with other species of tree shrew (Lim, 1967).

Ecology and behaviour: Pen-tailed tree shrews feed mainly on insects and fruit, although in captivity they may also accept meat. They are largely nocturnal in habit (the only tree shrew which displays this habit) and are excellent climbers, using the elongate tail for support and balance. They nest in holes in tree trunks or branches 20–30m high. The nest consists of a simple structure of dried leaves, twigs and fibres of soft wood. They generally move about in pairs but as many as four animals have been found together in a single nest.

Pen-tailed tree shrews are seldom seen and have not been studied in any great detail.

3.4 Captive Breeding of Tree Shrews

Tree shrews have long been a favourite exhibit for many zoological gardens. Many species have already been successfully maintained and bred in captivity. In fact, much of what is known about the social behaviour and breeding patterns of tree shrews has been the result of captive observations. This technique clearly has a great deal to offer and a more detailed review is required of current knowledge on this subject.

In the meantime, however, husbandry techniques are known to be well advanced for a number of species, particularly *Tupaia glis* and *T. minor*. The Philippine tree shrew (*Urogale everetti*) has also been successfully bred in several zoos. If conditions permit, it is recommended that further efforts should be undertaken to breed and manage some of the more threatened species in national collections, particularly *T. nicobarica, T. chrysogaster, T. longipes, T. palawensis* and *Urogale everetti*, using the experience already gained with the captive management of the more common species.

As with other species, however, captive propagation should only be considered as one component of the conservation strategy for any given species. Habitat protection, as well as maintenance and enhancement of the local environment should always remain the priority actions in these instances.

3.5 Conservation Requirements for Tree Shrews – A Summary

Tree shrews are forest-dwelling species, occupying a wide range of niches within different forest habitats. More versatile and seemingly more adaptable than terrestrial-dwelling species, many species of tree shrew appear to be able to adapt quite readily to secondary forest. Some even appear to be tolerant of, and adaptable to, habitat alterations which would effect most other forest-dwelling species. Supporting evidence for this comes from the fact that many tree shrews have been recorded from logged and secondary forests as well as rural gardens which mimic forest in structure. But no investigation has yet been made of the direct impact of forest clearance or associated disturbances on these species, nor of the length of time and conditions required for recolonisation. Given the present state of habitat destruction in many parts of south-east Asia, this type of investigation should be a priority for further actions and surveys.

In some parts of south-east Asia, however, tree shrews are already known to be seriously threatened by human encroachment and the destruction of their natural habitat. Where this occurs, additional pressures such as hunting for food and sport, can add considerably to the pressure on these and other native, endemic species. There is therefore a need for much more information to be obtained before a realistic overview of this situation can be presented.

From the preceding species accounts, it is clear that the ecology, conservation requirements and status of many species is still inadequately known. The most urgent among these are the golden-bellied (*Tupaia chrysogaster*), Nicobar (*T. nicobarica*), Palawan (*T. palawensis*), Bornean (*T. longipes*), Philippine (*Urogale everetti*) and Madras (*Anathana elliotti*) tree shrews. In addition, information is generally thought to be inadequate for a number of other species, particularly the ruddy (*Tupaia splendidula*), northern (*T. belangeri*), painted (*T. picta*), striped (*T. dorsalis*) and northern smooth-tailed (*Dendrogale murina*) tree shrews.

Habitat conservation is critical for tree shrews. Research has shown that some species occupy considerable home ranges: Kawamichi and Kawamichi (1979, 1982) have shown that *T. glis*, for example, has an average home range size of 10,174m^2 (for males) and 8809m^2 for females. Field observations and trapping results also suggest that many species exist at relatively low densities: data for *T. glis* show a varied pattern of 2–5/ha in Malaysia or 6–12/ha in Thailand (Langham, 1982; Lekagul and McNeely, 1977; Sorenson, 1970, 1974), while Dans (1993) has shown that *T. palawanensis* occurs at a local density of 1.6–3.2/ha. Payne (pers. comm.) reports that striped tree shrews (*T. dorsalis*) naturally exist at extremely low densities, although there are no indications that this species occupies any particular specialised habitat or niche. The mountain (*T. montana*) and smooth-tailed (*Dendrogale* spp.) tree shrews, although restricted in distribution, occur at locally high population densities in montane habitats which are subject to little disturbance (Payne, pers. comm.).

In view of the many uncertainties surrounding the ecological status, distribution and ability of these species to adapt to secondary habitats there is, therefore, an urgent need to promote further action for a large number of these species. Particular aspects which need attention include details of the basic ecology, home range size, feeding priorities, habitat preferences, breeding behaviour and population density of most species. This is best addressed in the first instance through additional field studies. Particular attention should be given to surveys of known, or soon-to-be-established, protected areas so that any management recommendations might be easily incorporated within broader activities for that location.

In the meantime, and on the basis of available data, a number of priority species have been identified which should receive attention. Recommended actions include:

- Detailed field surveys to determine the range, ecology and status of *Tupaia nicobarica*. Information from these surveys should form the basis of a recovery/management plan if judged necessary. Further recommendations should also be made to protect the habitat of this species.

- An investigation of the status, distribution and ecology of *T. longipes* on the islands off East Malaysia. Particular attention should be given to identifying existing or potential threats to this species.

- Field surveys to determine the distribution and status of *Tupaia chrysogaster* in Indonesia.

- Field surveys to determine the range and status of *Tupaia palawanensis* in the Philippines.

- An investigation of the status and range of *Anathana elliotti* in India, including a preliminary evaluation of threats to this species in view of the increasing loss of forest cover in this region.

- An evaluation of the status and habitat requirements of *Urogale everetti* in the Philippines.

- Detailed surveys of the adaptability of key species to habitat alteration and other disturbances.

- An investigation of the causes of the low population density of the striped tree shrew (*T. dorsalis*).

- An attempt to map the distribution and habitat preferences of the ruddy (*T. splendidula*) and painted (*T. picta*) tree shrews.

- Additional field surveys to obtain a more accurate reflection of species' and subspecies' distribution and population densities.

- Identification and evaluation of wider threats to all species and subspecies through factors such as hunting.

Chapter 4
Conservation Action Plan

4.1 Introduction

Conservation of Eurasian insectivores and tree shrews is only likely to succeed if it is fully integrated into the broader issues of environmental management and sustainable development. However, insectivores and tree shrews are generally not regarded as being the most charismatic mammals, and rarely feature as prominent species in national or regional conservation action plans. Few have any recognisable economic or social value, which prevents them from being adopted as flagship species which could help increase conservation awareness among decision-makers and the general public.

At the same time, however, few of these species interfere with man's activities and have therefore not been branded with a "pest" status, which could have serious implications on their numbers and/or distribution. Many species have therefore avoided pressure from direct persecution. However, where this has previously taken place, the results have been quite dramatic, leading for example in the cases of the European mole (*Talpa europaea*) and Russian desman (*Desmana moschata*) to local extinctions and seriously depleted population levels, respectively. The danger of over-exploitation should therefore not be ignored, even for seemingly common species.

The focus of world conservation issues is clearly not directed at the level of insectivores or tree shrews. Their specialised niches, small size, fossorial or aquatic lifestyles and frequently nocturnal habits have even resulted in many species being overlooked in scientific field surveys. Although major advances have now been made in research techniques, many species – particularly among the Soricidae and Talpidae – are too small, or require too specialised equipment to enable detailed field research to be carried out. Obtaining adequate information to determine the status (especially population size) and conservation requirements of certain species therefore remains a problem.

Many species of insectivores, particularly the shrews, exist at high densities. This phenomenon may vary according to the season as well as geographical location – a feature which needs to be accounted for in field surveys and management plans. Nonetheless, their impact upon the natural environment, whether through the large amount of invertebrates consumed, or in their role as prey species for many terrestrial and avian predators, is considerable. A better understanding of these species' ecology and their precise habitat requirements is, therefore, a critical issue for their long-term conservation, as well as for a balanced environment.

From the species accounts given in Chapter 3, it is clear that there is still a need for considerable research into the majority of species featured in this Action Plan. Information is sadly lacking for many species. At present, therefore, it is not possible to present a detailed account of the conservation status, habitat preferences, or even the actual or potential distribution of every species. For this reason, the information contained in the following sections is primarily intended to act as a guideline to future investigations. There is still a considerable amount of research required before all of these aspects can be properly addressed.

Two immediate types of action have been recognised to further the conservation of threatened Eurasian insectivores and tree shrews: habitat conservation and additional field surveys to determine species' distribution, abundance, ecology and conservation requirements (if any). Both activities are essential and should, wherever possible, be carried out in unison.

4.2 Conservation of Insectivores and Tree Shrews – Habitat Conservation

Many of the species covered in this Action Plan are forest-dwelling animals. Any loss of forest habitat, either through conversion to agricultural land, loss or disturbance as a result of logging for timber, clearance for fuelwood or charcoal, or the burning of undergrowth to encourage better grazing for livestock, are likely to have a considerable impact on insectivores and tree shrews. The rate at which deforestation is taking place, particularly in South and south-east Asia where such high concentrations of biodiversity occur, are therefore of particular concern. Many of these countries are priority sites for the conservation of insectivores and tree shrews. Indonesia, for example, may be losing as much as 12,000km^2 of forest each year, while the Philippines, a land which in its natural state was believed to be almost entirely forested, has lost four-fifths of its forest cover (Collins *et al.*, 1991).

Deforestation often leads to habitat fragmentation, as patches of forest are cleared in such a manner that they are no longer connected with each other. This reduces the possibility of migration or dispersal taking place, and may also result in unnaturally high population densities for the available food supplies. Of longer-term importance is the probability of inbreeding and loss of genetic variation. Combined, these features make isolated populations extremely vulnerable to extirpation through demographic instability or catastrophic events such as disease.

Similar results may arise from human interference with wetlands. The construction of a dam for purposes of hydro-electric power, agricultural irrigation, or as a water reservoir can have a major effect on aquatic and semi-aquatic species both above and below the construction site. The ranges of certain species can be divided into separate parts, but probably the most important changes are related to ensuing interruptions of the natural water flow. Stream drift, a biological phenomenon exhibited by many aquatic insects and other invertebrates, is an essential part of the daily rhythm of life in most streams and rivers, as many different organisms collectively rise from the stream bed and float downstream with the current. The timing of such events are often linked with peak feeding time amongst fish. The activity patterns of some semi-aquatic insectivores, such as the Pyrenean desman, may also be tied in with this event (Stone, 1987b). Stream drift, together with the transportation of sediment and other nutrients, are often disrupted by the construction of dams and barrages, resulting in an irregular supply of foodstuffs for many species in these food chains.

Other phenomena associated with the construction of dams include localised changes of temperature regimes, and a transformation of the landscape above the dam, in particular. Flooding caused by large dams and reservoirs can result in considerable habitat alterations, changing small rivers and streams into deep flooded valleys, many of which are unsuitable for the smaller insectivores. Such changes may also lead to the development of new floral and aquatic communities, which may not be favourable to species associated with fast-flowing water conditions.

Another major threat to semi-aquatic species worldwide is freshwater pollution. Rivers and streams are frequently used a convenient dumping grounds for a wide range of agricultural, municipal, industrial and other wastes, many of which have not been properly treated to remove toxic wastes. One of the most immediate effects of such pollutants is a decrease in the oxygen level in the water, a feature often associated with dense algal blooms. When this happens, many species die as a result of suffocation or starvation as their food supplies are rapidly destroyed.

Environmental problems related to both water pollution and acid rain are increasing worldwide. Both of these subjects are potential threats to insectivores because of their reliance on soil invertebrates as a primary food source.

Conserving the habitat in a near-pristine condition is, without doubt, the only practical means of conserving tree shrews and insectivores – species that have a wide range of specific needs. Any habitat which is set aside for either of these taxa – woodland, grassland, mountain streams or rainforest – will undoubtedly be of major benefit to a much wider range of other small mammals, invertebrates and plants. Many countries have already taken measures to conserve representative parts of their natural heritage through the establishment of national parks and other protected areas.

Protected areas are now viewed as essential in representing parts of a country's natural heritage if it is to be protected in the long term. Many species, even those with disjunct ranges, may benefit from part of their range being included within a national park, nature reserve, or similar protected area. The Pyrenean desman (*Galemys pyrenaicus*) or Russian desman (*Desmana moschata*), for example, benefit from the latter form of protection, although several additional sites have been established specifically for the Russian species. Likewise the recently described dwarf gymnure (*Hylomys parvus*) from Mt Kerinci, West Sumatra, will undoubtedly benefit from the establishment and wise management of the Kerinci-Seblat National Park.

The value of establishing protected areas such as these is that by protecting a vital habitat such as a headwater stream, montane forest, or floodplain with associate marshlands, a multitude of other species associated with freshwater conditions will also thrive. Increased protection and careful management can therefore result in a significantly increased level of biological diversity as well as considerable socio-economic benefits for people living in the vicinity of such regions.

A number of species, especially those with restricted ranges, are unlikely to occur in existing protected areas (see following national accounts for examples) and more effort is needed to ensure that adequate precautionary measures are take to protect these species. Wherever adequate survey data have been obtained to support further action for conserving a species, results should be reported to relevant authorities and every effort made to include the habitat of this species within the country's existing system of protected areas.

4.3 Conservation of Insectivores and Tree Shrews – Field Surveys

Any reader of this Action Plan will become immediately aware that there is an urgent need for a wide range of field surveys on the majority of species covered. Field surveys have a number of important roles to play in the present context as they:

- help identify areas of particularly high biological diversity;

- enable recommendations to be made for the conservation of such sites;

- are the only means of identifying the location, population size and basic ecology of isolated and fragmented populations, in what may well be high risk habitats; and

- are of considerable importance in helping determine to what extent threatened or potentially threatened species of Insectivora and Scandentia are represented in existing protected areas.

Only when an up-to-date status report of each species is available can the conservation status of individual species be evaluated and necessary actions taken. Thankfully, major advances have been made in identifying national and international sites of particularly high biodiversity and, in many of the countries featured in the following sections, considerable progress has already been made in establishing and managing a range of protected areas in many high priority sites, many of which are of indirect benefit to insectivores and tree shrews.

Within the present context, the following section presents a series of individual country reports that are intended as a first attempt to describe, identify and assess the status and particular conservation needs of threatened species, as recognised through the new system of categorisation proposed by IUCN (IUCN, 1995) and adopted in this Action Plan. In addition to drawing attention to these priority species, the following accounts also list a number of surveys or other recommendations for other species whose current conservation status is still uncertain.

The recommended actions for future work on these species is, therefore, a working combination of field surveys (including some scientific research where the need has been identified) and habitat conservation. Together these activities would play a vital role in ensuring that conservation needs have been identified and acted upon in time to save certain species from extinction.

As the following national overviews demonstrate there is now an urgent need for both activities, especially, but by no means exclusively, in the shrinking forests of south-east Asia – China, India, Indonesia, Malaysia and the Philippines, in particular. Also at risk are monospecific species with restricted distributions, and small island-dwelling species, particularly those of the Indonesian archipelago, Japan and the Philippines, but also others such as those on Hainan Island (China) and the Andaman and Nicobar Islands (India). Aquatic species are always likely to pose a conservation problem and special long-term protection measures should be taken to ensure that vital freshwater resources are adequately protected not only for these species, but also on account of the essential roles that freshwater supplies play in our daily lives.

As this is the first stage of evaluating and proposing a series of recommendations for further action, the status of each species will be under periodic review. Thus, when new information is available on a particular species, this will be applied to the IUCN criteria which may result in a reclassification of a certain species' status. In this way, it is hoped that this work will encourage additional, new surveys and that any such research will contribute to existing findings and contribute to the overall conservation of threatened species.

For convenience sake the following section is organised on a country-by-country basis although, in reality, such boundaries are artificial for the species concerned. The reader is therefore also advised to also consult the regional accounts for neighbouring countries. At this stage of evaluating the status of these species, little attention has been given to the numerous subspecies of insectivores or tree shrews. It is hoped that more attention can be directed towards subspecies in future but, for the moment it is apparent that because of taxonomic confusion and lack of field data on the behaviour, ecology and conservation status of the majority of these species, this would not be a realistic undertaking at the present time.

4.4 Country Assessment of Status and Conservation Needs for Threatened Insectivores and Tree Shrews

AFGHANISTAN

Low Priority
- The eastern range of the shrew *Crocidura gueldenstaedti* may occur in Afghanistan, although

this cannot be determined on account of current taxonomic uncertainties. Scientific research is needed to clarify this situation.

- The status of *Crocidura zarudnyi*, recorded from Afghanistan, western Pakistan and south-east Iran needs to be determined.

- The distribution of the long-eared hedgehog (*Hemiechinus auritus*) needs investigation. This species is, however, widely distributed in western Asia.

- A subspecies of Brandt's hedgehog (*Paraechinus hypomelas hypomela*) has been recorded from Afghanistan, although nothing is known about its ecology or conservation status. Future surveys of small mammals should take this subspecies into account.

ANDORRA

Key Species
Pyrenean desman (*Galemys pyrenaicus*) (VU: B1 and 2c)[1]

High Priority
- Detailed surveys should be carried out to determine the presence and extent of this species' range in this small mountainous principality. Possible threats to this species should also be examined. Steps should be taken to protect riparian habitat and water quality wherever viable populations are identified. The need to undertake conservation action, such as establishing a protected area, should be re-examined when further details are known.

ARMENIA

Key Species
Crocidura armenica (DD)

Moderate Priority
- Field surveys should be conducted in Armenia to determine the range and conservation status of the shrew *Crocidura armenica*.

BANGLADESH

Medium Priority
- The insectivore fauna of Bangladesh is poorly known. Future surveys for insectivores should concentrate in the Sunderbans (the only remaining extensive low-lying tracts of forest in the country), as well as in the hillier regions of the south and south-east, notably the Chittagong Hills. The mangrove and moist deciduous forest occurs of the Madhupur Tract and northern frontier with Meghalaya should also be surveyed. Specific sites which should be examined include: Himchari and Madhupur National Parks; the Sunderbans, Rema-Kalenga, Pablakhali, Char Kukri-Mukri and Chuntai Wildlife Sanctuaries; the proposed Hazarikhil and Hail Haor Wildlife Sanctuaries; and the Teknaf Game Reserve in the south-east.

BELARUS

Key Species
Russian desman (*Desmana moschata*) (VU: B1 and 2c)

High Priority
- Additional surveys should be carried out to evaluate and monitor the population size and distribution of the Russian desman. Further details of this proposal are provided in Appendix I.

Moderate Priority
- Wider surveys should be conducted of the insectivore fauna with a view to inclusion in protected areas. The Belovezhskaya Pushcha National Park, the country's only national park, should be surveyed.

BHUTAN

Moderate Priority
- Although many genera of shrews are represented in Bhutan, current data are inadequate to evaluate the needs of any species. Further investigations are needed and should include surveys of the Royal Manas, Black Mountain and Jigme Dorji National Parks, and wildlife sanctuaries. Surveys should also identify and focus on other remaining patches of the southernmost belt of forest which has been almost completely cleared for human settlement (Mahat, 1985). This is especially important now given the rate of conversion of forest to agricultural lands, the high consumption of fuelwood and timber, shifting cultivation, overgrazing and encroachment pressure on existing protected areas.

BRUNEI

Moderate Priority
- The distribution of insectivores and tree shrews is poorly known. Brunei's importance lies in the relatively untouched forest habitats which could be

[1] see Box 2.1 for Key

important refuges for moonrats and other species; but their occurrence needs to be confirmed. Surveys are therefore recommended for the extensive mangrove and peat swamps, as well as inland sites such as Batu Apoi Forest Reserve and the Bukit Batu-Sungei area which is contiguous with Gunung Mulu National Park in Sarawak. Additional surveys should be conducted in the Labi and Ladan hills, parts of which are established as forest reserves.

CAMBODIA

Moderate Priority
- The insectivore fauna of Cambodia is poorly known, but the country is thought to host several species of local and regional interest. Priority sites for investigation include the Cardamom Mountain range (which rises to over 1500m), and the mixed savanna zone north of Tonlé Sap, leading to the borders of Laos and Thailand.

PEOPLE'S REPUBLIC OF CHINA

Key Species
Hylomys hainanensis (EN: B1 and 2c)
Hylomys sinensis (LR: NT)
Mesechinus hughi (VU: B1 and 2c)
Crocidura shantungensis (DD)
Crocidura horsfieldii wuchihensis (DD)
Blarinella wardi (LR: NT)
Sorex cansulus (CR: B1 and 2c)
Greater stripe-backed shrew (*Sorex cylindricauda*) (EN: B1 and 2c)
Sorex excelsus (DD)
Kozlov's shrew (*Sorex kozlovi*) (CR: B1 and 2c)
Sorex sinalis (VU: B1 and 2c)
Salenski's shrew (*Soriculus salenskii*) (CR: B1 and 2c)
Mogera insularis (LR: NT)
Uropsilis soricipes (EN: B1 and 2c)
Uropsilis investigator (EN: B1 and 2c)

High Priority
- Field surveys should investigate the distribution and status of the Gansu shrew (*Sorex cansulus*) as a first priority. Almost nothing is known about the status of this species.

- Almost nothing is known about *Sorex kozlovi*, a shrew known only from the type locality in Tibet. Future field surveys of this country should attempt to determine whether this species is still present and, if so, to prepare suitable follow-up surveys and recommendations.

- The status of Salenski's shrew (*Soriculus salenskii*), known only from the type locality in northern Sichuan, needs to be determined.

- The gymnure *Hylomys hainanensis* has only been recorded from rainforest and subtropical evergreen forest on Hainan Island. Its precise distribution and conservation status needs to be determined in view of the rate of habitat loss on the island.

- Among the soricidae, priority species which should be the focus of future surveys include the greater stripe-backed shrew (*Sorex cylindricauda*) which has only been recorded from montane forest in Central Sichuan at an altitude of about 3000m.

- Of the shrew moles (Genus *Uropsilus*), the status of *U. soricipes* and *U. investigator* are probably of most immediate concern, both being restricted to a small area of forest and alpine habitat in Sichuan and Yunnan, respectively, at altitudes of 1500–2700m. More information is urgently required on these species.

- The conservation status of *Mesechinus hughi*, known only from Shaanxi and Shanxi Provinces should be investigated.

- Another restricted species which requires attention is *Sorex sinalis*, known only from Shaanxi and southern Gansu Provinces.

- Similar details are also required for a number of less-threatened species such as *Hylomys sinensis*, *Blarinella wardii* and *Mogera insularis*. Insufficient data are currently available to evaluate the status of *Crocidura shantungensis* and *Sorex excelsus*, but it is likely that both are threatened. The conservation status of these two species should be reviewed when more information comes available from these surveys.

- A subspecies of Horsfield's shrew (*Crocidura horsfieldii wuchihensis*) is only known from Hainan Island, specifically the western slope of Wuchih Mountain. More details are required on the status and habitat requirements of this species.

Moderate Priority
- There is a great need for more detailed surveys of the distribution and status of insectivores in China, with particular attention being given to the following sites in southern China – the tropical rainforests of Hainan and the island of Macao. On the mainland, the Xishuangbanna Nature Reserve, China's most important protected area within the Indo-Malayan

Realm, should be thoroughly surveyed. A priority listing of sites in need of further surveys should be prepared with relevant experts. These should include, but not be confined to, existing and proposed protected areas.

- Field surveys are required to determine the distribution and status of *Uropsilis andersoni*, which has only been recorded from South Sichuan.

- The Gansu mole (*Scapanulus oweni*) is another monospecific genus represented in China. This species has only been recorded from South Gansu, East Qinghai, south-west Shaanxi and North Sichuan. It appears to prefer a montane habitat, having been recorded at altitudes between 2700–3000m. Only about six specimens exist in museums. Future surveys of small mammals in these provinces should attempt to determine the status of this species.

Low Priority
- The status of the Manchurian hedgehog (*Erinaceus amurensis*) needs to be determined. The Daurian hedgehog (*Mesechinus dauuricus*) appears to be widely distributed in China but, again, precise details on its distribution are unavailable at the present time.

- Recorded from south-west China, the distribution of the least shrew (*S. minutissimus*) in this country should be determined, as should that of the Tibetan shrew (*Sorex thibetanus*).

- The status and distribution of De Winton's shrew (*Soriculus hypsibius*), which appears to be restricted to two disjunct areas in China – Sichuan and South Shaanxi (Qinling Shan) and Hebei – needs to be determined. Its preferred habitat appears to be montane forest, but this needs verification.

- The status of *Soriculus lamula* in China also needs clarification. This species has been recorded from north-west Yunnan, Central Sichuan and South Gansu at about 2000–3000m. A single specimen has also been recorded from Fujian; future surveys should also focus on this region.

- Smith's shrew (*Soriculus smithi*) is known only from Central and West Sichuan and the Qinling Mountains, South Shaanxi, China; while *S. parca* reaches its northern limit in Yunnan and Sichuan, where it contacts the range of *S. smithi* (Hoffmann, 1986). Additional details of the distribution and status of both species should be obtained during future surveys of these regions.

- The status of the web-footed water shrew (*Nectogale elegans*), the only species in its genus, needs to be determined. Although this species occurs widely in Burma, Nepal and Tibet, there are no precise data on its distribution or conservation status. In China it has been recorded from the mountain streams in the south-west (South Shaanxi, Sichuan, Qinghai and Yunnan Provinces).

- The long-tailed mole (*Scaptonyx fusicaudus*), a monospecific genus, appears to be largely confined to China – Shaanxi (Wu and Liu, 1982), East Qinghai, Sichuan and Yunnan Provinces – although some records do exist from northern Myanmar. This genus is represented by just a few specimens in study collections (Walker, 1991) and its status (and perhaps taxonomy) requires urgent attention. This species has been recorded within an altitudinal range of 2100–4100m. A single subspecies, *S.f. affinis* Thomas, has been described from north-west Yunnan. It too requires immediate investigation.

- The northern tree shrew (*Tupaia belangeri*) has been recorded from Guangxi, Yunnan and southern Sichuan Province, as well as Hainan Island, but its status and range limits are uncertain. Although this species is widely dispersed through Thailand, Burma, Bangladesh and East Nepal, more precise details should be gathered on its range in China.

CRETE

Key Species
Crocidura zimmermanni (VU: B1 and 2c)

High Priority
- Some surveys have already been carried out to map the distribution of this restricted island species. Future efforts should be directed to determining the habitat requirements, population parameters and conservation threats to this species. A detailed management plan might be required if this species' habitat is found to be a limiting factor.

CYPRUS

Low Priority
- Little information is available about insectivores on this island. Future surveys should include an evaluation of the long-eared hedgehog (*Hemiechinus auritus*) population. The Troodos Mountains in the extreme south should be the subject of an extensive survey for small mammals.

EGYPT

Key Species
Crocidura floweri (EN: B1 and 2c)
Crocidura religiosa (DD)

High Priority
- The conservation status and ecology of the shrews *Crocidura floweri* and *C. religiosa* should be investigated in future field surveys of small mammals.

FRANCE

Key Species
Pyrenean desman (*Galemys pyrenaicus*) (VU: B1 and 2c)

High Priority
- A detailed survey of the Pyrenean desman in the Pyrenees is a top priority (see also Appendix II) as this species occupies a restricted niche which is easily perturbed by human interference. Priorities should be to obtain an indication of the population size, as well as investigate in more detail the precise habitat requirements of this species. Future surveys should also help determine where additional protected areas in the Pyrenees might be established to protect this and other semi-aquatic, species.

Low Priority
- The westernmost extension of the range of *Suncus etruscus*, one of the smallest known mammals, is in the reed beds of the Camargue, south-east France. Surveys are required to determine the range and status of this species in France.

INDIA

Key species
Crocidura jenkinsii (CR: B1 and 2c)
Crocidura andamanensis (EN: B1 and 2c)
Andaman shrew (*Crocidura hispida*) (EN: B1 and 2c)
Nicobar shrew (*Crocidura nicobarica*) (EN: B1 and 2c)
Crocidura pergrisea (VU: B1 and 2c)
Suncus dayi (EN: B1 and 2c)
Suncus montanus (VU: B1 and 2c)
Nicobar tree shrew (*Tupaia nicobarica*) (EN: B1 and 2c)
Tupaia nicobarica surda

High Priority
- The status of the critically endangered *Crocidura jenkinsii*, known only from the type locality, South Andaman Island, needs immediate investigation. Recommendations for protection of this species' habitat should be prepared upon the findings of this survey.

- Other species of the Andaman Islands which also require urgent investigation include *Crocidura andamanensis*, which has also only been recorded from South Andaman Island: nothing is known about its ecology or status. Likewise, the status of the Andaman shrew (*Crocidura hispida*), known only from the Middle Andaman Island, is equally unclear.

- Another restricted island shrew species, the Nicobar Shrew (*Crocidura nicobarica*) has only been recorded from the Great Nicobar Island in the Bay of Bengal. As a priority species it, too, should be the subject of an immediate survey.

- The shrew, *Suncus dayi*, a poorly known species has only been recorded from southern India. Nothing is known about the ecology or geographical distribution of this species, which should be the subject of a thorough field survey.

- A second species of the same genus, *S. montanus* has been recorded from the Nilgiri and Palni Hills and may occur in other parts of southern India (as well as Sri Lanka). It, too, requires urgent action to determine its status and basic conservation requirements.

- The Nicobar tree shrew (*Tupaia nicobarica*) occurs on both Great and Little Nicobar Islands. A subspecies, *T.n. surda*, has been described from Little Nicobar Island on the basis of a duller, less yellow colour. The status of both species is of particular concern in view of the current rate of loss of forest habitat in these regions as agricultural activities continue to spread. Surveys to determine the exact status of both species should incorporate a detailed examination of their preferred habitat, together with an evaluation of these areas vis-a-vis existing or proposed protected areas.

- Detailed conservation recommendations should be prepared on the basis of the findings of the above surveys, especially for the Andaman and Nicobar Islands.

Moderate Priority
- More specific information is required on the distribution, status and ecology of *Crocidura pergrisea* – at present known only from Baltistan, Shigar and Skora Loomba: these records should be confirmed with new surveys which should also be extended over a wider area to determine the precise limitations of this species' range.

- The status of the Indian hedgehog (*Paraechinus micropus*), known only from north-west India and

Pakistan, is poorly known. Its range is apparently divided into two segments: from the Indus (including the western bank of the delta) through Rajasthan and Gujarat, south to at least Bombay and Pune and, to the east, towards Agra. The second area is in Tamil Nadu, southern India where a distinct species, *P. nudiventris*, may occur (Corbet, 1992). Confirmation of this range is required, as well as more precise details on the ecological requirements of this species.

- The distribution and conservation status of *Hemiechinus collaris*, adapted to living in arid conditions in north-western India, also needs investigation.

- Verification of the taxonomic status and distribution of a subspecies of the Sikkim large-clawed shrew (*Soriculus nigrescens caurinus*) in India needs to be determined.

- The status of Horsfield's Shrew (*Crocidura horsfieldii*) also needs to be clarified. At present it has only been recorded from the southern regions.

- Anderson's Shrew (*Suncus stoliczkanus*) has been described from Peninsular India – from Madras north to Rajasthan and the Indus – as well as from neighbouring Nepal (Abe, 1982), and south-west Yunnan (China) (Wang and Liu, 1980). Its status in India needs clarification.

- The Madras Tree Shrew (*Anathana elliotti*), a monotypic species, only occurs in the forested regions of Peninsular India, as far north as Bahar, in the east, and the Satpura Hills (south-west Madhya Pradesh) in the west. The habits of this species are not well known but are probably similar to those of *Tupaia* and *Urogale*. The status of this endemic species needs to be investigated.

Low Priority
- The status of the northern tree shrew (*Tupaia belangeri*) in Assam should be determined, although this is a widely distributed species in neighbouring countries.

INDONESIA

Key species
Hylomys parvus (CR: B1 and 2c)
Chimmarogale sumatrana (CR: B1 and 2c)
Crocidura beccarii (EN: B1 and 2c)
Crocidura minuta (DD)
Crocidura orientalis (VU: D2)
Crocidura paradoxura (EN: B1 and 2c)

Crocidura tenuis (VU: B1 and 2c)
Suncus mertensi (CR: B1 and 2c)
Tupaia chrysogaster (VU: B1 and 2c)

High Priority
- Enhanced protection of Mt Kerinci and the surrounding Kerinci-Seblat forests to protect the habitat of the recently described *Hylomys parvus*. Further surveys should attempt to estimate the range, status and habitat preferences of this species. Measurement should be made of range overlap and interactions with *H. suillus* on Kerinci.

- Field surveys in the Padang Highlands, West Sumatra, to confirm the presence of the Sumatran water shrew (*Chimmarogale sumatrana*). In the first instance, an attempt should be made to gauge the extent of this species' distribution as well as obtaining some basic information on the ecological requirements of this specialised shrew.

- Survey of the Island of Flores and adjacent islands for *Suncus mertensi*, known only from the type specimen; particular attention should be given to surveying existing protected areas such as the Ruteng Nature Reserve and Lewotoli Protection Forest.

- Conservation of montane forest on Mt Singgalang, West Sumatra, the only known location for *Crocidura beccarii* and *C. paradoxura*. Field surveys are required to obtain basic information on the ecology, habitat requirements and conservation status of these species. Additional surveys should be conducted in outlying forest sites to investigate the occurrence of these species at other sites and to better define the limits of their distribution.

- Conduct a field survey on the island of Timor to confirm the presence of *Crocidura tenuis* – currently known from two damaged specimens. Surveys should include, but not be restricted to, existing protected areas such as Dataran Bena Hunting Park, Mt Timau Forest reserve and Mt Mutis Protection Forest. Habitat requirements for this species should be determined as quickly as possible and appropriate steps taken to ensure that key locations are adequately protected.

- Conservation of montane forest on Mt Gede-Pangrango, West Java, for *Crocidura orientalis*. The distribution of this species should be determined through appropriate field projects. Any such surveys should also be used to determine whether *C. paradoxura* occurs at this site (as suggested by Corbet, 1992).

- The range, status and ecological requirements of the golden-bellied tree shrew (*Tupaia chrysogaster*), an endemic species of the Mentawai islands, needs to be determined.

- Data is urgently required on the status and distribution of *Crocidura minuta*, currently known only from East Java. The IUCN status of this species should be corrected once additional information becomes known.

Moderate Priority
- Field surveys to determine the range of *Crocidura neglecta*, which has been recorded from Sumatra (perhaps including Mt Kerinci), East Java, Sumba, Flores and perhaps Ambon (Moluccas) (Jenkins, 1982).

- Field surveys to determine the distribution of *Crocidura elongata* and *C. lea*, both of which have only been recorded from lowland and montane forest in North and Central Sulawesi. Surveys in Sulawesi should also record the distribution of *Crocidura levicula* (collected from lowland and montane forest on Central and south-eastern Sulawesi); *Crocidura nigriceps* (known from lowland forest on North and Central Sulawesi, including Lembeh Island. (A single subspecies, *C.n. lipari*, has been described on the basis of its larger size.); and *Crocidura rhoditis* (another forest species recorded from North, Central and south-western Sulawesi). Many of these surveys could be carried out in conjunction with ongoing conservation efforts at the following sites of particular interest: Lore Lindu National Park, Mt Rantemario, Morowali Nature Reserve, and Bogani Nani Wartabone National Park.

- Among the Tupaiidae, the distribution and status of the Javan tree shrew (*Tupaia javanica*), known generally from Bali, Java, West Sumatra and the island of Nias needs to be clarified.

- The status of the painted tree shrew (*T. picta*) in eastern Kalimantan needs to be investigated.

- The status of two subspecies of the ruddy tree shrew (*T. splendidula*) – *T.s. splendidula* from mainland Borneo and Natuna Island, and *T.s. lucida* from Laut Island and North Natuna Island – needs to be examined.

- General surveys of the insectivoran fauna should be carried out in a number of high priority reserves to determine the presence of other insectivore species and evaluate their overall status. Among the main sites of interest are the Kerinci-Seblat National Park, Gunung Leuser National Park and the Mentawai Islands (Sumatra); Bromo Tengger Semeru, G. Gede-Pangrango, G. Halimun and Ujung Kulon National Parks (Java); Tanjung Puting, Kayan Mentarang and Kutai National Parks (Kalimantan); Bogani Nani Wartabone and Lore Lindu National Parks and Morowali Nature Reserve (Sulawesi); as well as isolated islands in the Lesser Sundas and Moluccas.

Low Priority
- Determine the range of *Crocidura monticola* in Indonesia. This species has been recorded from Peninsular Thailand (Davison, 1984), Malaysia, Borneo, Java and Ambon (Indonesia). It may also exist on the islands of Sumbawa, Lombok, Sumba, Flores, Komodo, Obi and Timor, but all of these need confirmation (Jenkins, 1982).

- The status of the moonrat (*Echinosorex gymnurus*) on Sumatra should be determined.

ISLAMIC REPUBLIC OF IRAN

Key species
Persian mole (*Talpa streeti*) (CR: B1 and 2c)
Crocidura susiana (EN: B1 and 2c)

High Priority
- The status of the Persian mole (*Talpa streeti*) known only from the type locality, Hezer Darrak in Kurdistan Province, north-west Iran, needs to be determined through appropriate field surveys. The conservation status of this species needs to be determined, existing and potential threats identified, and specific management recommendations drawn up.

- An investigation of the ecology and conservation status of the shrew *Crocidura susiana*. The taxonomic status of this species also needs to be re-examined.

Low Priority
- Further surveys are needed to determine the status of most insectivore species. These should concentrate in the first instance on the country's varied steppe forests, as well as broadleaved temperate forests of northern regions.

- The occurrence and precise distribution of *Crocidura zarudnyi* (known only from western Pakistan, Afghanistan and south-east Iran) needs clarification.

ITALY

Key Species
Crocidura russula cossyrensis (VU: D2)

Moderate Priority
- *Crocidura russula cossyrensis* (Vogel *et al.*, 1992) has only been recorded from Pantelleria Island. Future surveys should investigate the distribution and conservation status of this subspecies, and identify existing or potential threats.

Low Priority
- More precise details should be obtained through field surveys to determine the range of the Appenine shrew (*Sorex samniticus*), found only in southern Italy. The extent of range overlap between this species and more common *S. araneus* needs to be determined in the Appenines.

- More precise information needs to be obtained on the geographical limits of certain species, especially the fossorial *Talpa romana*.

JAPAN

Key Species
Sorex sadonis (EN: B1 and 2c)
Sorex hosonoi (VU: B1 and 2c)
Crocidura orii (EN: B1 and 2c)
Euroscaptor mizura (VU: B1 and 2c)
Nesoscaptor uchidai (EN: B1 and 2c)
Mogera etigo (EN: B1 and 2c)
Mogera tokudae (EN: B1 and 2c)

High Priority
- An appraisal is urgently needed of the status of the Sado shrew (*Sorex sadonis*), known only from Sado Island.

- A survey should be carried out on the general insectivoran fauna of the Ryukyu islands, with particular attention given to determining the status of *Crocidura orii*, a endangered species restricted to these islands.

- An investigation of the status and ecology of *Nesoscaptor uchiai* on the Ryukyu and Senkaku islands.

- Further surveys of the status of *Mogera etigo* and *M. tokudae* should attempt to evaluate the population size of these species, as well as determine the degree of threat facing these endangered species.

- Field surveys should be conducted in and around the montane regions of Central Honshu, the only known location of the Azumi shrew (*Sorex hosonoi*). Recommendations should be prepared to protect the remaining habitat of this species.

- The Japanese mountain mole (*Euroscaptor mizura*) is confined to just a few isolated montane areas on Honshu, where it has been recorded from forest and alpine grassland. More information is urgently required on the distribution and ecological requirements of this species. When known, appropriate conservation measures should be introduced to protect the habitat of this species.

Moderate Priority
- The Japanese white-toothed shrew (*Crocidura dsinezumi*) occurs mainly on Honshu and the islands of Kyushu, Shikoku, Yakushima, Tanegashima, Amamioshima, Oki and Okinoshima. It has also been reported from Quelpart Island (South Korea). More information is required on its range and habitat requirements.

- One species of Asiatic water shrew, the endemic *Chimarrogale platycephala*, has been reported from Honshu, Skikoku and Kyushu. Further details are required on the ecology and conservation requirements (if any) of this species.

- Both members of the genus *Urotrichus* – the lesser Japanese shrew-mole *(U. pilirostris)* and the greater Japanese shrew mole *(U. talpoides)* – occur only in Japan where they are restricted to montane habitats, particularly regions of coniferous forest up to 2000m. *U. pilirostris* is only found on the islands of Honshu, Shikoku and Kyushu, preferring montane coniferous forest, while *U. talpoides* is restricted to forest and grassland habitats on the islands of Honshu, Shikou, Kyushu, Dogo and North Tsushima Island. Both species are reported to be plentiful but do not occur on the plains. Future surveys should determine their distribution and abundance.

- Further surveys should be carried out to determine the precise range of other insectivores in the islands. Many of the islands of the Ryukyu Archipelago, which are included in the Indo-Malayan biogeographic realm, still support important broadleaved evergreen forests. Although many endemic species are known from these islands, little attention has focused on the small mammal fauna to date. Particular attention should be given to the following islands: Iriomote, Amami, and Okinawa Kaigan, which have already been recognised as priority sites by IUCN (MacKinnon and MacKinnon, 1986).

KAZAKHSTAN

Key Species
Russian desman (*Desmana moschata*) (VU: B1 and 2c)

High Priority
- Additional surveys should be carried out to evaluate and monitor the population size and distribution of the Russian desman. Further details are provided in Appendix I.

Medium Priority
- Field surveys should be carried out to determine the range and conservation needs of other insectivores in this region, with particular attention being given to existing protected areas such as Bayanaul'sky National Park.

- The range and status of the piebald shrew *Diplomesodon pulchellum*, the only species within this genus, occurs in desert environments of southern Kazakhstan (Kirghiz Steppes) and Turkmenistan. Future surveys in this region should take account of this species.

DEMOCRATIC PEOPLE'S REPUBLIC OF KOREA

Low Priority
- The status of the Manchurian (Amur) hedgehog (*Erinaceus amurensis*), which occurs in forest and grassland, should be determined.

- The precise status of the giant shrew (*Sorex mirabilis*), which has been recorded from the Ussuri region of eastern Siberia and North Korea, needs to be examined.

- The status of the slender shrew (*Sorex gracillimus*), whose range stretches from Siberia from the southern shore of the sea of Okhotsk to North Korea and probably Manchuria, should also be determined.

- The southern limit of the Ussuri white-toothed shrew (*Crocidura lasiura*) occurs in Korea, although the limits of its range are unclear.

- Among fossorial species, the status and distribution of the greater mole (*Talpa robusta*) needs to be examined.

- Surveys of Mt Kumgang National Park and Mt Paekdu Biosphere Reserve, the country's only protected areas.

THE REPUBLIC OF SOUTH KOREA

Low Priority
- The status of the Manchurian (Amur) hedgehog (*Erinaceus amurensis*) should be determined.

- The distribution pattern of the greater mole (*Talpa robusta*) should be clarified.

- The status of the Japanese white-toothed shrew (*Crocidura dsinezumi*) on Quelpart Island needs to be confirmed, and information obtained on its status and habitat requirements.

- General surveys should be carried out in lowland and montane forest sites to determine the presence and range of insectivores. Particular attention should be given to existing national parks, most of which are established around mountains.

LAO PEOPLE'S DEMOCRATIC REPUBLIC

Medium Priority
- Little information is available concerning the distribution and/or conservation status of Insectivora in Laos. Priority sites for conservation are the proposed reserves of Attapeu, Lai Leng, Luang Prabang and Xé Kaman (MacKinnon and MacKinnon, 1986) as well as the Bolovens Plateau. Increasing population growth, rural settlement and clearance of natural vegetation has taken a heavy toll on the country's natural forest cover. The most extensive forests are now confined primarily to the southern and central regions, deforestation having been most severe in the north (IUCN, 1991a).

MALAYSIA

Key Species
Chimarrogale hantu (CR: B1 and 2c)
Chimarrogale phaeura (EN: B1 and 2c)
Crocidura malayana (EN: B1 and 2c)
Suncus ater (CR: B1 and 2c)
Suncus hosei (VU: B1 and 2c)
Dendrogale melanura (VU: B1 and 2c)
Tupaia longipes (EN: B1 and 2c)

High Priority
- The Malayan water shrew (*Chimarrogale hantu*), known only from specimens collected at the Ulu Langat Forest Reserve, Selangor State, West Malaysia should be the subject of an immediate investigation. Field surveys should attempt to locate this species within the reserve before carrying out similar activities elsewhere in the region. Management recommendations may need to be prepared for this species and its habitat.

- The black shrew (*Suncus ater*) is known only from a single specimen trapped from montane forest on Mt

Kinabalu, Sabah (1700m). Additional surveys should be undertaken at this important site of biodiversity to check whether this species still survives.

- The Bornean Water Shrew (*Chimarrogale phaeura*) occurs in northern Borneo (where it is only known from Mt Kinabalu and Mt Trus Madi). A semi-aquatic species, it is dependent on unpolluted mountain streams at an altitudinal range of 460–1700m. Further surveys should be carried out to obtain more information on the extent of its range as well as its habitat requirements and existing or potential threats.

- More precise details are needed on the exact range of *Crocidura malayana* in Peninsular Malaysia. The presence of this species also needs to be confirmed on offshore islands.

- Another species of shrew, *Suncus hosei* has been reported from montane forest in Sabah and northern Sarawak, although it is not clear whether this represents a true species. Further surveys and additional taxonomic research should be undertaken to determine the status of this species.

- More information is required on the status and distribution of the Bornean smooth-tailed tree shrew (*Dendrogale melanura*) in Sabah and Sarawak.

- Field surveys should be conducted on the islands of Siantan, Riabu and Jimaja to determine the status and requirements of the tree shrew *Tupaia longipes*.

Moderate Priority
- Two subspecies of the moonrat (*Echinosorex gymnurus*), a widely distributed species through south-east Asia, have been described in Malaysia: *E.g. albus* from the eastern and southern regions of Borneo and the Kelabit uplands, as well as on Sumatra and Peninsular Malaysia (Corbet, 1992; Payne *et al.*, 1985); *E.g. candidus* from the western side of Borneo from P. Labuan south to at least Kuching region (Payne *et al.*, 1985). Future surveys should attempt to obtain more information on the precise range of these subspecies.

- In parts of its range *Echinosorex* is hunted as a food source and may also be trapped unintentionally in traps set for other animals. The extent of such offtake is unknown but may have an important effect at the local level; this activity needs to be investigated.

- Data are required on the distribution and status of the lesser gymnure (*Hylomys suillus*), known from Peninsular Malaysia, Sabah and Sarawak. The status of *H.s.dorsalis* on Borneo needs to be determined, as does the situation for *H.s. tionis* on Tioman Island.

- The status of *Crocidura monticola*, known from Peninsular Thailand (Davison, 1984), Malaysia, Java, Borneo and several smaller islands needs investigation.

- Malaysia is the most important centre for tree shrews with over 60% of the recognised taxa occurring. Although no species appears to be in immediate danger, the status of several subspecies should, however, be examined in more detail through additional surveys and scientific research. Particular attention should be given to:

- *Tupaia gracilis gracilis* from mainland Borneo and Banggi;

- *T. minor caedis* and *T.m. minor*;

- *T. montana baluensis* and *T.m. montana* from Sabah and Sarawak;

- *T. picta fuscior* Medway 1965 (as described from Labuan Klambu, East Kalimantan); and *T.p. picta* (Payne *et al.*, 1985);

- *T. tana besara*; *T.t. speciosa*; *T.t. utara*; *T.t. nitida*; *T.t. kelabit*; *T.t. chrysura*; *T.t. paitana*; and *T.t. kretami* from Borneo and Sumatra;

- *Dendrogale melanura baluensis* (from Mt Kinabalu) and *D.m. melanura*; and

- *Ptilocercus lowi continensis* (described from near Kuala Lumpur, Peninsular Malaysia).

- Malaysia has an extensive and well-managed network of protected areas, many of which already offer protection to a number of insectivores and tree shrews. Priorities for future surveys in the region should include Endau Rompin, the Cameron highlands, Genting Highlands, Fraser's Hill and Kedah Peak, the north-west peak region and Taman Negara on Peninsular Malaysia. On Sabah, Kinabalu National Park has already been well surveyed but other areas including the Crocker Range and Ulu Padas should be surveyed in greater details. Priority sites for Sarawak include Bako National Park, the Kelabit Highlands, Gunung Mulu National Park and the Lambir hills.

MALTA

Low Priority
- The distribution of the Algerian hedgehog (*Erinaceus algirus*) needs to be determined (Malec and Storch, 1972).

- Surveys should also be conducted in remaining forests and grassland to determine the distribution of Soricidae, particularly *Crocidura sicula*.

PEOPLE'S REPUBLIC OF MONGOLIA

Low Priority
- A wide range of Insectivora are known to occur in Mongolia, especially steppe- and arid zone-dwelling species. This is especially true for the Soricidae (see Sokolov and Orlov, 1980). The conservation status of most cannot be determined at the present time. Future surveys of the small mammal fauna should be encouraged. Given current exploitation patterns and rates, few are likely to be threatened on a country-wide basis although some local extinctions could occur as a result of activities such as mining or deforestation.

MYANMAR

Key Species
Hylomys sinensis (LR: NT)
Blarinella wardii (LR: NT)

Moderate Priority
- The distribution of the lesser gymnure (*Hylomys sinensis*) should be determined through field surveys.

- The status of the southern short-tailed shrew (*Blarinella wardi*) should be examined in northern Myanmar. This species is also represented in Yunnan Province (China), where its status is equally unclear.

- A distinctive subspecies of the Sikkim large-clawed shrew (*Soriculus nigrescens radulus*) has been reported from Myanmar (Corbet, 1992). Nothing is known about its status. Further field surveys and taxonomic research are required to address this.

- One species of shrew mole, *Uropsilus gracilis*, has been recorded from northern Myanmar (where its range is probably an extension of the species' Chinese distribution). It is the only known record of this genus outside China; more precise information should be obtained for this species.

- The status of the long-tailed mole (*Scaptonyx fusicaudus*), recorded also from China, should be investigated as there are only few records of its occurrence in northern Myanmar.

Low Priority
- The distribution of the moonrat (*Echinosorex gymnurus*), the largest representative of the Insectivora in Myanmar, is inadequately known. Surveys should be conducted in forests to determine its distribution and relative abundance.

- Among the Soricidae, the status of the following species should be determined:

- lesser stripe-backed shrew (*Sorex bedfordiae*);

- Horsfield's Shrew (*Crocidura horsfieldii*);

- Hodgson's brown-toothed shrew (*Soriculus caudatus*);

- *Soriculus macrurus*;

- Indian long-tailed shrew (*Soriculus leucops*); and

- *Soriculus parca*.

All of these species have been recorded from northern regions, overlapping with ranges in China, but there are no precise records of the extent of their distribution, relative abundance or conservation status in Myanmar.

- The status of the Szechuan mole shrew (*Anurosorex squamipes*), a monotypic genus, from northern Myanmar needs to be examined, although this species is widely represented in other neighbouring countries.

- Another monotypic species with a restricted range in the mountains and highlands of North Myanmar and China, at an altitudinal range of 2000–3500m, is the Chinese short-tailed shrew (*Blarinella quadraticauda*). Again, precise details on this species are lacking; future surveys should take this species into account.

- The status and habitat requirements of the Himalayan Water Shrew (*Chimarrogale himalayica*), and Styan's Water Shrew (*C. styani*) in north and north-eastern Myanmar, respectively, needs to be determined. Any survey of this nature should also be extended to include an investigation of the distribution of the Tibetan water shrew (*Nectogale elegans*) – a monotypic genus – in northern regions.

- Among the obligate fossorial species, more detailed information should be obtained for *Parascaptor leucura*.

NEPAL

Key Species
Sorex excelsus (DD)

Moderate Priority
- The presence of the shrew *Sorex excelsus*, otherwise recorded only from Yunnan and Sichuan Provinces (China) needs to be determine in this country. Its conservation status should be reviewed when more information becomes available.

Low Priority
- Insectivores of particular concern are the lesser stripe-backed shrew (*Sorex bedfordiae*) and its subspecies, *S.b. fumeolus* and *S.b. gomphus*, which have been recorded from Nepal, North Myanmar and South Gansu, Sichuan, and West Yunnan (China) in montane forest at altitudes of 2100–4400m.

- A subspecies of Hodgson's brown-toothed shrew (*Soriculus caudatus soluensis*) has been identified from East Nepal, although nothing is known about its range or ecology.

- The range of *Soriculus macrurus*, which extends from Central Nepal to Sikkim, from north-west Myanmar and West and South Yunnan to Sichuan (China) and North Vietnam, should also be investigated.

- Another species sharing a similar geographical range, but with equally poor information on its distribution in Nepal, is *Soriculus leucopus*. Additional surveys are required.

- The status of a subspecies of the Sikkim large-clawed shrew (*Soriculus nigrescens centralis*) from Bouzini needs to be investigated.

- The status of Anderson's Shrew (*Suncus stoliczkanus*) in Nepal needs to be clarified.

- The elegant water shrew (*Nectogale elegans*), the only species in its genus, inhabits clean mountain streams at altitudes of about 900–2270m. Its distribution and status in Nepal is uncertain and should be determined through field surveys.

- Among fossorial insectivores, the status of the Himalayan Mole (*Euroscaptor micrura*) in eastern Nepal needs confirmation.

SULTANATE OF OMAN

Key Species
Crocidura dhofarensis (CR: B1 and 2c)

High Priority
- The shrew *Crocidura dhofarensis* (Hutterer and Harrison, 1988) is known only from the type locality at Khadrafi. More precise details are required on the range, ecology and conservation status on this species.

PAKISTAN

Low Priority
- Two subspecies of Brandt's hedgehog (*Paraechinus hypomelas*) have been described: *P.h. hypomela* from western Pakistan including Baluchistan and Afghanistan; and *P.h. jerdoni* from the Indus Valley. Although no data are available to judge the conservation status of these two subspecies it is felt that a field survey should be organised to investigate the situation.

- The status of *Crocidura zarudnyi*, which has been recorded in western Pakistan (as well as Afghanistan and south-east Iran), also needs to be investigated.

THE PHILIPPINES

Key Species
Podogymnura truei (EN: B1 and 2c)
Podogymnura aureospinula (EN: B1 and 2c)
Crocidura beatus (VU: B1 and 2c)
Crocidura grandis (EN: B1 and 2c)
Crocidura grayi (VU: B1 and 2c)
Crocidura mindorus (EN: B1 and 2c)
Crocidura negrina (CR: B1 and 2c)
Crocidura palawanensis (VU: B1 and 2c)
Tupaia palawanensis (VU: B1 and 2c)
Urogale everetti (VU: B1 and 2c)

High Priority
- Additional field surveys should be carried out to determine the status of the two species of *Podogymnura*, both of which are restricted to the southern Philippines. The Mindanao moonrat (*Podogymnura truei*) is known only from Mindanao, having been collected on Mt Apo at elevations of 1700–2100m, on the eastern slope of Mt McKinley from 1800–2300m and on Mt Katanglad at an elevation of 1600m. Much of the forest habitat which this species prefers has been destroyed by logging and shifting agriculture. The other species in this genus, *P. aureospinula*, is restricted to Dinagat Island and is

only known from a few specimens collected from logged dipterocarp forest. Conservation recommendations need to be prepared for both species once adequate baseline information on habitat requirements and threats has been compiled.

- A series of field projects are required to determine the distribution and conservation status of the following shrews, all endemic to the Philippines:

- *Crocidura beatus* (Mindanao, Leyte and Maripipi islands);

- *C. grandis* (Mindanao);

- *C. grayi* (Luzon and Mindoro);

- *C. mindorus* (Mindoro and Sibuyan islands);

- *C. negrina* (Negros Island); and

- *C. palawanensis* (recorded only from Palawan).

- The status of the Palawan tree shrew (*Tupaia palawanensis*) is unclear. Further studies need to be conducted on the ecology of this species in the wild.

- Conservation of montane forest on Dinagat, Siargao and Mindanao to conserve the monotypic Philippine tree shrew (*Urogale everetti*), and the wide range of other insectivores that share this important habitat.

Moderate Priority
- Conduct additional surveys of the insectivore and tree shrews in the following regions: the lowland rainforest of Samar and Sibuyan Islands; the Sierra Madre mountains, northern Luzon; Palawan; Mt Apo; Mt Giting-giting; and Mt Pulog. Appropriate conservation recommendations should be prepared for submission to higher authorities once necessary information has been gathered and where immediate threats are recognised.

PORTUGAL

Key Species
Pyrenean desman (*Galemys pyrenaicus*) (VU: B1 and 2c)

High Priority
- The only insectivore of known conservation concern is the Pyrenean desman (*Galemys pyrenaicus*) which is confined to unpolluted, fast-flowing streams of northern Portugal and also northern Spain and the Pyrenees Mountains. Further surveys are required in the first instance to determine more precisely the range and possible population density of this species, as well as to document the extent of threats. Additional details on this species are outlined in Appendix II.

Low Priority
- Future surveys of small mammals in Portugal should attempt to map the precise distribution of the Soricidae.

RUSSIA

Key Species
Paramushir shrew (*Sorex leucogaster*) (VU: B1 and 2c)
Russian desman (*Desmana moschata*) (VU: B1 and 2c)

High Priority
- The status and distribution of the Paramushir shrew (*Sorex leucogaster*), currently known only from Paramushir Island, south of Kamchatza Peninsula, needs to be examined.

- The Russian desman (*Desmana moschata*) is considered threatened largely as a result of excessive hunting for its fur in past decades, but also as a result of recent and continuing habitat alteration. Additional surveys should be carried out to evaluate and monitor the population size, distribution and current status of this species in West Russia. Further details are provided in Appendix I.

Medium Priority
- The status of the Kamchatka shrew (*Sorex camtschatica*), known only from the southern Kamchatka Peninsula needs to be investigated.

- Field surveys should be carried out to determine the range and conservation needs of other insectivores in this region.

SAUDI ARABIA

Low Priority
- The distribution and status of a number of hedgehogs is unclear. The hedgehog *Paraechinus hypomelas* is thought to be rare, confined to the highlands and southern regions of the peninsula. One subspecies *P.h. niger* is recognised in Saudi Arabia, while another, *P. aethiopicus albatus*, is known only from the type locality; this represents the eastern limit of the species. The status of *P.a. dorsalis* should also be investigated.

- The status of *Crocidura arabica*, a species more widely represented in Yemen and Oman, should be investigated to determine the extent of its range.

SPAIN

Key Species
Crocidura canariensis (VU: B1 and 2c)
Pyrenean desman (*Galemys pyrenaicus*) (VU: B1 and 2c)
Crocidura osorio (VU: D2)

High Priority
- More detailed information is required on the ecology and conservation status of *Crocidura canariensis*, known only from the East Canary islands. This species is currently protected under Spanish law (Hutterer, 1993).

- The only insectivore of known conservation concern on the mainland is the Pyrenean desman (*Galemys pyrenaicus*) which is confined to unpolluted, fast-flowing streams of northern Portugal, northern Spain and the Pyrenees Mountains. Further details of the requirements for this species are outlined in Appendix II.

- Further surveys are required to determine the ecological requirements and precise status of *Crocidura osorio*, known only from patches of montane forest on Grand Canary Island.

Low Priority
- The shrew *Sorex granarius* is known to have a limited distribution in the mountains of Central Spain. Its conservation status is unknown and future field surveys should attempt to determine its distribution.

- General surveys of the insectivore fauna should be conducted in the Baetic mountains and sub-Baetic mountains, Spanish Pyrenees and northern Cantabrians, in particular.

SRI LANKA

Key Species
Crocidura miya (EN: B1 and 2c)
Feroculus feroculus (EN: B1 and 2c)
Solisorex pearsoni (EN: B1 and 2c)
Suncus fellowsgordoni (EN: B1 and 2c)
Suncus zeylanicus (EN: B1 and 2c)
Suncus montanus (VU: B1 and 2c)

High Priority
- Field surveys should attempt to determine the exact distribution and status of the Sri Lankan long-tailed shrew (*Crocidura miya*), known only from the central highlands.

- One of the main priorities for future field surveys in Sri Lanka should be for Kelaart's long-clawed shrew (*Feroculus feroculus*), one of the least-known insectivore species. Fewer than 10 specimens of this montane forest-dwelling species have been collected. Specific data should be obtained on the habitat requirements of this species, as well as an evaluation of threats. Conservation recommendations should be drawn up on the basis of data obtained.

- Pearson's long-clawed shrew (*Solisorex pearsoni*), a monospecific genus, is endemic to the forests of the central highlands. Its status is unknown. Additional field surveys should obtain data necessary to evaluate the need for further conservation action.

- Other shrew species that need to be examined include *Suncus fellowsgordoni* (known only from the central highlands); *S. montanus*, recorded only from rainforests in Central and southern Sri Lanka and southern India; and *S. zeylanicus* from Central and southern Sri Lanka (in moist forest between 150m and 1000m in Sabaragamuwa and Central Provinces). Conservation recommendations should be prepared on the basis of field data.

Moderate Priority
- Conduct additional surveys for insectivores in the following reserves, known for their high levels of biodiversity: the forests of Kottawa, Hinidumkanda, Kanneliya and Gilimale, which are currently unprotected (Guntilleke and Guntilleke, 1990), as well as the Sinharaja Forest Reserve, Peak Wilderness Sanctuary, Horton Plains National Park and the Knuckles Mountain Range.

Low Priority
- The status of Horsfield's shrew (*Crocidura horsfieldii*) should be determined. This species may be quite widespread throughout the island and not of immediate conservation concern.

TAIWAN

Key Species
Mogera insularis (LR: NT)

Moderate Priority
- Field surveys should attempt to map the distribution of *Mogera insularis*, known only from Taiwan and Hainan Island (China). A preliminary evaluation should also be carried out on existing and potential threats to this species.

- Another species of particular concern is *Soriculus fumidus*, an endemic species to the island. This species appears to be restricted to forest and dwarf bamboo stands at an altitude of 1000–3200m. More details are required on the distribution, population size and extent of threat to this species. Conservation recommendations should also be prepared.

- Future survey priorities should include Kenting National Park, well known for its botanical diversity, Ta-Wu Mountain Preserve and Yushan National Park. Problems such as illegal hunting and trapping are unlikely to have an impact on insectivore populations, but more serious threats include clearance of forest and land reclamation for agricultural purposes, roads and other developments such as hydro-electric programmes. The extent of such threats should be evaluated once more information is available on the range and distribution of insectivore species on the island.

Low Priority
- The Szechuan mole shrew (*Anourosorex squamipes*), a monospecific genus, occurs on Taiwan, although no further details are available on its distribution. Although this species is widely distributed on the Asian mainland, its status on Taiwan should also be examined.

- The shrew *Crocidura gueldenstaedti* may also occur in Taiwan (if *C. hosletti* is conspecific). Its existence should be confirmed.

- The range of the grey shrew (*Crocidura attenuata*) should also be determined through additional field surveys.

THAILAND

Key Species
Suncus malayanus (DD)

Moderate Priority
- The conservation status and distribution of the shrew *Suncus malayanus* should be examined through field studies. Its IUCN status should be re-examined once sufficient data have been obtained.

- The distribution of *Soriculus parca* in northern Thailand needs to be examined.

- The Szechuan mole shrew (*Anurosorex squamipes*), the only species in its genus, also occurs in North Thailand but precise details of its range are lacking.

- The status of *Crocidura monticola* from Peninsular Thailand (Davison, 1984) also needs clarification.

- Among fossorial species, the sole representative in Thailand is the Himalayan Mole (*Euroscaptor micrura*). Although this is a relatively widespread species occurring from the Himalayas west probably as far as eastern Nepal, it is only known from isolated localities in Assam, Thailand, Laos, Vietnam and Malaysia. Its status and distribution in Thailand should therefore be determined.

- Of the five species of tree shrew recorded in Thailand, the only species of concern is the northern smooth-tailed tree shrew (*Dendrogale murina*). Its range is restricted to southern Indochina – specifically eastern Thailand, southern Vietnam and Cambodia. Future field studies should attempt to determine the extent of its range and identify possible threats in Thailand.

- Additional field surveys of insectivores and tree shrews should be carried out in the country's wide range of existing protected areas.

TURKEY

Low Priority
- Renowned for its botanical diversity, few surveys have been conducted on Turkey's small mammal fauna in general, or insectivores, in particular. Although no threatened species have been identified, surveys are needed to investigate the species diversity (especially amongst the Soricidae), distribution and abundance. These could concentrate on already known centres of endemism, especially near Erzincan, Erzurum, the mountains south of Lake Van and on chalk outcrops near Cankiri and Sivas. Surveys in the Mediterranean region should focus on the Lyciam and Cilician Taurus, while others should be conducted in the north-east region.

THE UKRAINE

Key Species
Russian desman (*Desmana moschata*) (VU: B1 and 2c)

High Priority
- Additional surveys should be carried out to evaluate and monitor the population size and distribution of the *Desmana moschata*. Further details are provided in Appendix I.

Medium Priority
- Field surveys should be carried out to determine the range and conservation needs of other insectivores in

this region. Recommendations for additional protected areas should be drawn up if the results of these surveys suggest that some species are threatened.

Low Priority
- The distribution of the Ukrainian shrew (*Sorex volnuchini*), known from South Ukraine and the Caucasus and, possibly Turkey and North Iran, should be determined.

THE SOCIALIST REPUBLIC OF VIETNAM

Key Species
Euroscaptor parvidens (CR: B1 and 2c)
Hylomys sinensis (LR: NT)

High Priority
- The status of *Euroscaptor parvidens* should be determined through future field surveys. This species is only known from the type locality, Di Linh, Vietnam and Rakho, on the border with China.

Moderate Priority
- The precise distribution of *Hylomys sinensis* in northern Vietnam (where it has been recorded from cool, moist forest) should be determined. This species also occurs in montane areas of Sichuan and Yunnan (China) and adjacent parts of Myanmar.

Low Priority
- The status of *Soriculus macrurus*, *S. parca* and the Indian long-tailed shrew (*S. leucops*), which also occur in neighbouring countries, is unclear. Future surveys should attempt to obtain more detailed information on these species.

- The distribution of the Szechuan mole shrew (*Anourosorex squamipes*), a monotypic species which also occurs in neighbouring countries, should be determined.

- The distribution of the Himalayan water shrew (*Chimarrogale himalayica*) needs to be examined in much more detail.

- Among the fossorial species, the Himalayan mole (*Euroscaptor micrura*) is only known from a few isolated records. The status of *E. longirostris* is also unclear.

- Additional general surveys of the insectivore fauna are needed to obtain a more detailed overview of the level of biodiversity represented in Vietnam. Many of Vietnam's species are also represented in neighbouring countries and are probably in no immediate danger. Local extinctions are, however, possible although it is currently not possible to determine, or even anticipate, these from current data. Continuing loss of habitat occurs as a result of agricultural clearance, forest fires, collection of fuelwood, timber gathering and urban expansion. There is therefore an urgent need for new surveys of all small mammals from this region. Future surveys should concentrate on determining the insectivore fauna of existing and proposed protected areas.

4.4 Priorities for Action

It is apparent from the information so far presented that additional population surveys and considerably more information on the basic ecological requirements are required for a very large number of species. From the preceding species and country accounts it is possible to draw up a preliminary list of species which are recognised as threatened and which should be addressed, collectively or individually, as a matter of great importance. This information is presented in Table 4.1 according to taxonomic affiliations. For convenience sake, this same information is also presented according to conservation status, which may be of assistance in determining a priority ranking of those species included in this Action Plan (see Appendix IV).

Application of IUCN's new criteria for identifying threatened species (as used to generate the details in Table 4.1) has clearly separated out a number of different categories of threat currently facing Eurasian insectivores and tree shrews. Included in this table are 13 "Critically Endangered" species, 30 "Endangered", 22 "Vulnerable", three "Near Threatened" and six species which have been categorised as "Data Deficient". From this, it is recommended that priority actions should first address those species listed as "Critically Endangered" and "Endangered". Actions that are urgently required include *both* habitat protection and field surveys to determine the extent of individual species' ranges, some indication of the population size, and a preliminary evaluation of threats to respective species.

For species listed as "Vulnerable" and the lower categories of threat, it is recommended that additional field surveys should be set in motion to determine the distribution, habitat requirements, population and existing and potential threats to each species. Additional scientific research may be necessary to clarify the taxonomic status of selected species. Particular attention should be given in the first instance to isolated species, semi-aquatic species and others found only on small islands, since these are most at risk.

Table 4.1. Threatened species of insectivores and tree shrews

Genus	Species	Conservation status	Distribution
Family Erinaceidae			
Mesechinus	M. hughi	Vulnerable	Shaanxi and Shanxi Provinces, China
Hylomys	H. hainanensis	Endangered	Hainan Island, China
	H. parvus	Critically Endangered	West Sumatra, Indonesia
	H. sinensis	Lower Risk (NT)	Sichuan and Yunnan (China); adjacent Myanmar and North Vietnam
Podogymnura	P. aureospinula	Endangered	Dinagat Island, the Philippines
	P. truei	Endangered	Mindanao, the Philippines
Family Soricidae			
Crocidura	C. andamensis	Endangered	South Andaman Island, Indian Ocean
	C. armenica	Data Deficient	Armenia, Caucasus
	C. beatus	Vulnerable	Mindanao, Leyte, Maripipi, the Philippines
	C. beccarii	Endangered	West Sumatra, Indonesia
	C. canariensis	Vulnerable	East Canary islands
	C. dhofarensis	Critically Endangered	Oman
	C. floweri	Endangered	Egypt
	C. grandis	Endangered	Mindanao, the Philippines
	C. grayi	Vulnerable	Luzon and Mindoro, the Philippines
	C. hispida	Endangered	Middle Andaman Island, Indian Ocean
	C. jenkinsii	Critically Endangered	South Andaman Island, Indian Ocean
	C. malayana	Endangered	West Peninsular Malaysia
	C. mindorus	Endangered	Mindoro, the Philippines
	C. minuta	Data Deficient	East Java, Indonesia
	C. miya	Endangered	Central highlands, Sri Lanka
	C. negrina	Critically Endangered	Negros Island, the Philippines
	C. nicobarica	Endangered	Great Nicobar Island, Bay of Bengal
	C. orientalis	Vulnerable (D2)	West Java, Indonesia
	C. orii	Endangered	Ryukyu Islands
	C. osorio	Vulnerable (D2)	Grand Canary Island
	C. palawanensis	Vulnerable	Palawan, the Philippines
	C. paradoxura	Endangered	West Sumatra, Indonesia
	C. pergrisea	Vulnerable	Kashmir
	C. religiosa	Data Deficient	Egypt
	C. shantungensis	Data Deficient	Shandong Province, China
	C. susiana	Endangered	South-west Iran
	C. tenuis	Vulnerable	Timor Island, Indonesia
	C. zimmermanni	Vulnerable	Crete
Feroculus	F. feroculus	Endangered	Sri Lanka
Solisorex	S. pearsoni	Endangered	Sri Lanka
Suncus	S. ater	Critically Endangered	Mt Kinabalu, Borneo
	S. dayi	Endangered	Southern India
	S. fellowsgordoni	Endangered	Sri Lanka
	S. hosei	Vulnerable	Sabah and northern Sarawak
	S. malayanus	Data Deficient	Peninsular Thailand
	S. mertensi	Critically Endangered	Flores Island, Indonesia
	S. montanus	Vulnerable	Sri Lanka; southern India
	S. zeylanicus	Endangered	Sri Lanka

Table 4.1 (cont.). Threatened species of insectivores and tree shrews

Genus	Species	Conservation status	Distribution
Blarinella	B. wardi	Lower Risk (NT)	Northern Myanmar; Yunnan (China)
Chimmarogale	C. hantu	Critically Endangered	Malay Peninsula
	C. phaeura	Endangered	Northern Borneo
	C. sumatrana	Critically Endangered	West Sumatra, Indonesia
Sorex	S. cansulus	Critically Endangered	Gansu, China
	S. cylindricauda	Endangered	Sichuan, China
	S. excelsus	Data Deficient	Yunnan and Sichuan (China); possibly Nepal
	S. hosonoi	Vulnerable	Central Honshu, Japan
	S. kozlovi	Critically Endangered	Tibet
	S. leucogaster	Vulnerable	Paramushir Island (Russia)
	S. sadonis	Endangered	Sado Island, Japan
	S. sinalis	Vulnerable	West-central China
Soriculus	S. salenskii	Critically Endangered	Northern Sichuan, China

Family Talpidae

Genus	Species	Conservation status	Distribution
Desmana	D. moschata	Vulnerable	West Russia, Belarus, The Ukraine and Kazakhstan
Galemys	G. pyrenaicus	Vulnerable	Pyrenees Mountains (France and Spain), North and Central Spain, North Portugal
Euroscaptor	E. mizura	Vulnerable	Honshu, Japan
	E. parvidens	Critically Endangered	Vietnam
Mogera	M. etigo	Endangered	Honshu, Japan
	M. insularis	Lower Risk (NT)	Taiwan; Hainan Island, China
	M. tokudae	Endangered	Sado Island and Honshu, Japan
Nesoscaptor	N. uchidai	Endangered	Ryukyu and Senkaku islands, Japan
Talpa	T. streeti	Critically Endangered	North-west Iran
Uropsilis	U. investigator	Endangered	Yunnan, China
	U. soricipes	Endangered	Central Sichuan, China

Family Tupaiidae

Genus	Species	Conservation status	Distribution
Dendrogale	D. melanura	Vulnerable	Borneo
Tupaia	T. chrysogaster	Vulnerable	Indonesia
	T. longipes	Endangered	Siantan, Riabu and Jimaja islands, Malaysia
	T. nicobarica	Endangered	Nicobar Islands, Indian Ocean
	T. palawanensis	Vulnerable	Palawan, Balabac, Culion, Busuanga and Cuyo (the Philippines)
Urogale	U. everetti	Vulnerable	Mindanao, Dinagat and Siargao islands, the Philippines

Attention should also be given to improving our knowledge on the status and distribution of all monospecific genera, 14 of which feature in this Action Plan. Just six of these genera are included in the priority selection shown in Table 4.1, the remainder having been identified – on the basis of current knowledge – to not be under immediate threat. The situation of the remaining members of this group (*Atelerix, Echinosorex, Anourosorex, Diplomesodon, Nectogale, Parascaptor, Scapanulus, Scaptochirus*) should not be overlooked in future surveys.

With the introduction of this new system of categorisation by IUCN, ITSES will attempt to develop and find support for activities of the type discussed above to investigate the status of these threatened species. To coordinate these activities, a separate database of threatened species will be developed to monitor developments and progress on all fronts. It is proposed that individual species accounts, of the nature of those presented in Appendix I and II, be prepared for all threatened species. These reports would be prepared by members of ITSES, whose responsibility it would be to ensure that all relevant details are recorded and that efforts to investigate the status of individual species are fully coordinated among researchers.

Appendix I

Action Plan for the Russian Desman: Belarus, Kazakhstan, Russia, The Ukraine

An Evaluation of the Status of the Russian Desman (*Desmana moschata*)

Background – The Conservation Need

The Russian desman (*Desmana moschata*), an endemic species of eastern Europe, is a specialised semi-aquatic insectivore. During the Pleistocene this species ranged widely across Europe in a broad band from southern Britain in the west to the Caspian Sea in the east. Since then, the range of the species has shrunk considerably: by historical times, it was already confined to the southern part of European Russia in the basin of a few major river valleys.

The preferred habitat of this species is low-lying floodplains, where it lives in lakes, ponds and rivers with little or weak water currents. Deep-water lakes with a 2–3m detritus/sediment base which supports a rich benthic fauna, and a well developed littoral zone are essential habitat requirements for this species. Most of their food is captured underwater. When resting, desmans sleep in a long burrow in the river bank.

This species was relatively abundant until the late nineteenth century when hunting pressures for its valuable

The habitat of the Russian desman (*Desmana moschata*). (Photo by Gennady Khakhin)

fur eliminated and decimated many populations. At the turn of the twentieth century about 20,000 skins were being processed annually (Grzimek, 1975). Desman hunting was banned in 1929. Attempts to increase the population level once again consisted of releasing some 10,000 animals to the wild (Khakhin, 1993). Lack of management and poor protection, together with some licensed hunting of desmans in 1940, meant that the overall number of animals did not increase. Hunting was again banned in 1957. In the intervening years, however, the number of free-living desmans has continued to decline; estimates suggest that the population is now about 40,000 animals.

Compared to the Pyrenean desman (see Appendix II and preceding species accounts), the Russian species is relatively easy to survey on account of the large holes it makes in river banks or on lake shores. Desman populations are reported from the basins of the following rivers: Volga (23,000 animals), Don (10,000–12,000 animals), Dneiper (2000–3000), Ural (2000), and the Uj and Tobal rivers (2000–2500 animals).

Apart from hunting, the main reasons for the decline of this species – both in terms of overall number and distribution – have been drainage of lakes, water pollution, creation of impoundments for electricity, etc., clearance of riparian vegetation, fishing with nets and excessive grazing of floodplains.

As it has been the subject of many intensive studies, a considerable amount of information is available locally on this species and its conservation requirements.

The Conservation Requirements

This programme is primarily concerned with strengthening ongoing conservation initiatives for the protection of the Russian desman in its natural state.

1. Conservation Status and Requirements of the Russian Desman

The only member of its genus, the desman is listed as "Vulnerable" by IUCN. It is included in the Red Data book for Russia, The Ukraine, Belarus and Kazakhstan.

2. Past and Current Threats

Formerly widespread and more numerous, the desman population has been seriously reduced as a result of over-hunting (see earlier species account). Recognition of the impact of such hunting levels led to the introduction of a hunting ban in 1920. Limited hunting was again permitted under license in 1940, with a total ban again being introduced in 1957 (Khakhin, 1993).

One of the threats to the Russian desman (*Desmana moschata*) is accidental capture in fishermen's nets. (Photo by Gennady Khakhin)

At present, the main threat to this species is habitat alteration and degradation. Drainage of lakes and marginal wetlands, construction of hydro-electric dams, increased levels of fishing with nets, clearance of riparian vegetation, and competition for breeding sites with introduced nutria and muskrats are among the main threats to this species. As lakes become more isolated on floodplains, a high rate of mortality may be experienced from predation of migrating desmans.

3. Evaluating the Need for Captive Propagation

The Russian desman has already been the subject of intensive captive breeding efforts and re-introduction schemes. Unfortunately, limited information is available on the background and techniques used in this process. Khakhin (1993) reports that captive breeding trials were carried out from 1920 (when the first hunting ban was introduced) to 1929. A total of 10,000 animals were released in 1929, although the source of these animals is not clear. Some animals were released in parts of their former range; others were released in new areas of suitable habitat. However, a combination of poor management and lack of protection failed to extract the maximum benefits from this initiative, and the population level continued to decrease.

A certain amount of information is already available on the practical requirements of maintaining a captive population of desmans (Khakhin, 1993). The Hopior State Reserve has been established for this reason. Again, however, few details are available on this issue. It is therefore not feasible to discuss the needs or merits of captive breeding at the present time, although more details should be obtained on the way in which this operation was carried out. Ongoing field work, supported with research on semi-captive trials and wider environmental surveys should, in the coming years, enable a better appreciation to be made of the need for captive breeding.

4. Priority Actions

A range of actions have already been taken to protect this species in its natural state. Several protected areas were established in the 1920s and 1930s, specifically to protect the Russian desman.

One of the main reserves which has been established for desmans is the Oka State Reserve (Riazan region) which, with a floodplain of more than 30km wide, constitutes one of the largest floodplains in Europe. About 18% of this area is lacustrine, some of which are temporary water bodies. During the spring floods, the area is inundated for up to one and half months each year. During this time, desmans can easily migrate to other areas. Temporary nests are made in trees and bushes. When the water level recedes, however, desmans may become isolated in small lakes, many of which may dry up during the course of the summer, forcing the desmans to migrate overland in search of an alternative home.

Establishing and maintaining an adequate series of nature reserves would therefore appear to be a critical stage in the long-term protection of this species. Khakhin (1993) reported that five nature reserves and 80 refuges had already been established to protect this species. There is, however, an urgent need to re-evaluate the situation and status of these, and other sites.

In particular, the following priority actions should be carried out:

Half-submerged live traps ensure that no harm comes to captured Russian desmans (*Desmana moschata*). (Photo by Gennady Khakhin)

- Habitat inventory: in view of the many changes taking place in this region at the current time, there is an urgent requirement for a new habitat inventory to be conducted.

- On the basis of these results, draw up a comprehensive set of recommendations to support the establishment of additional reserves, and the reclassification of existing sites, if necessary. Necessary legislation should be prepared and enforced.

- Conduct an evaluation of the human influence in these habitats. Attempts should be made to minimise or avoid conflicts between human needs and those of desmans, wherever possible.

- On the basis of these findings, draw up a conservation programme for the Russian desman and its habitat submit it to the relevant regional and state bodies for approval. This programme should set out a 5–10 year conservation programme, including management policies for the reserves (patrolling, management of water regimes, use by local people, etc.), research priorities and procedures, etc.

- Develop accurate methods of carrying out population surveys and habitat evaluations, and train researchers in their application.

- Strengthen and continue ongoing research efforts to enable the compilation of a wider baseline of field data.

- Continue research into captive breeding efforts. Animal husbandry techniques should be re-examined and improved, if necessary. The objectives of the captive breeding programme should be clearly stated within the overall Management Programme. The removal of wild animals for captive breeding purposes should be avoided.

- Make plans for future re-introductions on a secure scientific basis. Past and ongoing research efforts have already provided a great deal of essential information which should be incorporated in any such plans.

- Develop a regional conservation and environmental awareness programme to greatly increase the long-term survival chances of the desman and its habitat. Local presentations at schools and other meetings should take advantage of the desman and its unique situation to introduce the concept of conservation, while demonstrating the many benefits that this could have for other species, as well as man.

Appendix II

Action Plan for the Pyrenean Desman: France, Portugal and Spain

An Evaluation of the Status of the Pyrenean Desman (*Galemys pyrenaicus*)

The Conservation Need

A relict species from the Tertiary era, the Pyrenean desman, *Galemys pyrenaicus*, is a specialised semi-aquatic insectivore adapted to living in cold, montane streams. The historical distribution of this species is unclear, but most remains have been recorded from the Pyrenees and their foothills. Their fossil remains are, however, easily overlooked and the species may once have been more widespread that we can currently imagine. Even today, its exact distribution is not well known but is generally referred to as the Pyrenean Mountain range (both the French and Spanish sides), Central and northern Spain, and northern Portugal.

The Pyrenean desman is adapted for life in cold streams. Its altitudinal range spans from 400–2500m, although it is rarely found at low elevations. According to Richard (1985), the ideal habitat for this species is a torrential stream flowing through meadows, bordered by boulders and dry-stone walls and lined with open-canopied trees, such as ash and alder, which do not cast too intense a shade across the water. The stones of the stream bank, together with exposed roots of riparian vegetation provide good shelter and access for nest construction under the bank.

The desmans forages mainly at night, feeding on a wide range of crustaceans and insect larvae, including mayfly, stonefly and caddisfly larvae. A highly secretive species, the desman is often known to local people (in France it is referred to as the "ratte à trompette") and is occasionally caught in fish nets.

The Conservation Requirements

The Pyrenean desman is a highly specialised insectivore which requires a number of precise environmental factors to ensure its survival in the wild. This conservation programme is primarily concerned with investigating the ecology and conservation requirements of the desman under free-living conditions.

The Pyrenean desman has been the subject of several studies (Richard and Vallette-Viallaird (1969), Richard (1985), and Stone (1985, 1986, 1987a,b). More recently, the Pyrenean and Russian desmans have been the subject of an international conference, the proceedings of which may be found in Queiroz (1993). There exists, therefore, a considerable amount of information on the basic habitat requirements and ecology of this species.

1. Conservation Status and Requirements of the Pyrenean Desman

The species is listed as 'Vulnerable' by IUCN and receives local protection in both France and Spain.

Parts of its range are situated within the Parc National des Pyrénées Occidentales and the Parque Nacional de Covadona. It is possible that the species also occurs in the Parque Nacional de Aiguas y Lago de San Mauricio and the Parque Nacional de Ordesa as well.

2. Past and Current Threats

The desman is now becoming increasingly threatened over much of its range. Alterations to the natural hydrological pattern are one of the main causes of interference. The construction of dams for hydro-electric power alters the natural (seasonal and daily) water flow rate below the dam, while impoundments on the upstream side also modify the natural environment. Diversion of streams to form canals also destroy many natural streams. Clearance of riparian vegetation removes important cover from river banks, leading to erosion and loss of nest sites, as well as a loss of nutrient input to the aquatic ecosystem through allochthonous materials (leaves, grasses, etc.), an important food source for many aquatic invertebrates. Clearance of vegetation also leads to increased evaporation and opens the river up to further development, some of which may be destructive to the freshwater ecosystem. Increasing agricultural development in many areas leads to run-off of pesticides and inorganic fertilizers which result in pollution, increased biological oxygen demand (BOD) and eutrophication. Of major concern is increasing fragmentation of the species' range as a result of habitat alteration. This can have a serious impact on local desman populations if they are not able to disperse to other suitable streams.

In the past, desmans have also been collected as a novelty species, using electro-fishing techniques. The extent of this practice is not known. Another, more recent and perhaps more serious threat to desmans comes in the form of introduced mink (*Mustela vison*) which have escaped from fur farms in northern Iberia. The versatility and adaptability of this species in the wild is cause for concern, particularly when dealing with such a restricted species as the Pyrenean desman. Again, however, the extent (if any) of this problem has not yet been investigated.

3. Evaluating the Need for Captive Propagation

The Pyrenean desman has never been an abundant species, at least in recent times. Now that the natural environment of this species is increasingly threatened, it is not clear whether wild populations will remain viable in future years. One of the main objectives of this evaluation project should be to determine whether a captive breeding programme is required.

Sufficient knowledge is already available on the desmans' feeding and activity regimes. This, together, with basic information on the species' ecology should permit a number of semi-captive trials to be carried out, initially using a series of cages built on the river. Special attention should be given to ensuring continuous access to clean, flowing water as well as access to dry land where the animal can feed, groom and rest.

Maintaining desmans in captivity should be carried out with the intention of gathering more information on animals husbandry needs and to attempt breeding these animals under captive conditions. However, too much emphasis should not be placed on this part of the programme, considering that the best (and cheapest) option is always the safeguarding of the natural environment. Captive breeding for desmans should therefore not be viewed as a vital part of the conservation programme; it should be undertaken in order to acquire essential details on husbandry and reproduction, but should always be seen as a final emergency reaction.

4. Priority Actions

The Pyrenean desman serves as an indicator species for assessing the status of the freshwater ecosystem in montane conditions. It has become increasingly obvious that this species is facing a wide range of threats, none of which has been quantitatively evaluated to date. Few direct actions have been taken to protect this species in the wild. There is therefore an urgent need to rectify this situation, before the species is removed from too much of its range.

Further investigations needed include:

- Wider habitat surveys in all parts of the desman's range. Quantitative, as well as qualitative, data should be obtained to identify the most important elements of satisfactory habitat for this species.

- Population surveys to be carried out throughout the desman's range to determine its exact distribution. An index of population level should also be determined.

- Completion of an assessment of current and potential threats to this species and its habitat. This should include an evaluation of existing or planned developments which might impact the freshwater ecosystem.

- Field research to extend current knowledge of this species, particularly by examining the social and spatial behaviour of the desman throughout the year, as well as its reproductive behaviour. Other details which need to be examined in more detail are seasonal movements and dispersal, population dynamics, predation and competition.

- On the basis of the habitat and population surveys, preparation of a management plan for the Pyrenean desman and its habitat. Exchange of information between researchers should be actively encouraged to further advances in field investigations.

- For each country, identification of where the Pyrenean desman and its habitat might benefit from the establishment of protected areas. Where desman populations already occur within protected areas, special measures should be encouraged to determine how best to meet the specific needs of conserving this species in the long-term.

- A public awareness programme to help people better understand the need for prudent conservation measures and also help them to identify how they might become involved with such activities. This should be a major component of the programme for the conservation of the Pyrenean desman because increasing development and tourism in the mountains is resulting in greater impacts on the environment.

Appendix III

Criteria for Critically Endangered, Endangered and Vulnerable Categories

Source: IUCN (1995)

CRITICALLY ENDANGERED (CR)

A taxon is Critically Endangered when it is facing an extremely high risk of extinction in the wild in the immediate future, as defined by any of the following criteria (A to E):

A. Population reduction in the form of either of the following:

1. An observed, estimated, inferred or suspected reduction of at least 80% over the last 10 years or three generations, which ever is longer, based on (and specifying) any of the following:
   (a) direct observation
   (b) an index of abundance appropriate for the taxon
   (c) a decline in area of occupancy, extent of occurrence and/or quality of habitat
   (d) actual or potential levels of exploitation
   (e) the effects of introduced taxa, hybridisation, pathogens, pollutants, competitors or parasites.

2. A reduction of at least 80%, projected or suspected to be met within the next 10 years or three generations, which ever is longer, based on (and specifying) any of (b), (c), (d) or (e) above.

B. Extent of occurrence estimated to be less than 100km² or area of occupancy estimated to be less than 10km², and estimates indicating any two of the following:

1. Severely fragmented or known to exist at only a single location.

2. Continuing decline, observed, inferred or projected, in any of the following:
   (a) extent of occurrence
   (b) area of occupancy
   (c) area, extent and/or quality of habitat

   (d) number of locations or sub-populations
   (e) number of mature individuals.

3. Extreme fluctuations in any of the following:
   (a) extent of occurrence
   (b) area of occupancy
   (c) number of locations or sub-populations
   (d) number of mature individuals

C. Population estimated to number less than 250 mature individuals and either:

1. An estimated continuing decline of at least 25% within three years or one generation, which ever is longer; or

2. A continuing decline, observed, projected, or inferred, in numbers of mature individuals and population structure in the form of either:
   (a) severely fragmented (i.e. no sub-population estimated to contain more than 50 mature individuals)
   (b) all individuals are in a single sub-population.

D. Population estimated to number less than 50 mature individuals.

E. Quantitative analysis showing the probability of extinction in the wild is at least 50% within 10 years or three generations, which ever is longer.

ENDANGERED (EN)

A taxon is Endangered when it is not Critically Endangered but is facing a very high risk of extinction in the wild in the near future, as defined by any of the following criteria (A to E):

A. Population reduction in the form of either of the following:

1. An observed, estimated, inferred or suspected reduction of at least 50% over the last 10 years or three generations, which ever is longer, based on (and specifying) any of the following:
   (a) direct observation
   (b) an index of abundance appropriate for the taxon
   (c) a decline in area of occupancy, extent of occurrence and/or quality of habitat
   (d) actual or potential levels of exploitation
   (e) the effects of introduced taxa, hybridisation, pathogens, pollutants, competitors or parasites.

2. A reduction of at least 50%, projected or suspected to be met within the next 10 years or three generations, which ever is longer, based on (and specifying) any of (b), (c), (d), or (e) above.

B. Extent of occurrence estimated to be less than 5000km² or area of occupancy estimated to be less than 500km², and estimates indicating any two of the following:

1. Severely fragmented or known to exist at no more than five locations.

2. Continuing decline, inferred, observed or projected, in any of the following:
   (a) extent of occurrence
   (b) area of occupancy
   (c) area, extent and/or quality of habitat
   (d) number of locations or sub-populations
   (e) number of mature individuals.

3. Extreme fluctuations in any of the following:
   (a) extent of occurrence
   (b) area of occupancy
   (c) number of locations or sub-populations
   (d) number of mature individuals

C. Population estimated to number less than 2500 mature individuals and either:

1. An estimated continuing decline of at least 20% within five years or two generations, which ever is longer; or

2. A continuing decline, observed, projected, or inferred, in numbers of mature individuals and population structure in the form of either:
   (a) severely fragmented (i.e. no sub-population estimated to contain more than 250 mature individuals);
   (b) all individuals are in a single sub-population.

D. Population estimated to number less than 250 mature individuals.

E. Quantitative analysis showing the probability of extinction in the wild is at least 20% within 20 years or five generations, which ever is longer.

VULNERABLE (VU)

A taxon is Vulnerable when it is not Critically Endangered or Endangered but is facing a high risk of extinction in the wild in the medium-term future, as defined by any of the following criteria (A to E):

A. Population reduction in the form of either of the following:

1. An observed, estimated, inferred or suspected reduction of at least 20% over the last 10 years or three generations, which ever is longer, based on (and specifying) any of the following:
   (a) direct observation
   (b) an index of abundance appropriate for the taxon
   (c) a decline in area of occupancy, extent of occurrence and/or quality of habitat
   (d) actual or potential levels of exploitation
   (e) the effects of introduced taxa, hybridisation, pathogens, pollutants, competitors or parasites.

2. A reduction of at least 20%, projected or suspected to be met within the next 10 years or three generations, which ever is the longer, based on (and specifying) any of (b), (c), (d) or (e) above.

B. Extent of occurrence estimated to be less than 20,000km² or area of occupancy estimated to be less than 2000km², and estimates indicating any two of the following:

1. Severely fragmented or known to exist at no more than 10 locations.

2. Continuing decline, inferred, observed or projected, in any of the following:
   (a) extent of occurrence
   (b) area of occupancy
   (c) area, extent and/or quality of habitat
   (d) number of locations or sub-populations
   (e) number of mature individuals.

3. Extreme fluctuations in any of the following:
   (a) extent of occurrence

(b) area of occupancy
(c) number of locations or sub-populations
(d) number of mature individuals

C. **Population estimated to number less than 10,000 mature individuals and either:**

1. An estimated continuing decline of at least 10% within 10 years or three generations, which ever is longer, or

2. A continuing decline, observed, projected, or inferred, in numbers of mature individuals and population structure in the form of either:
   (a) severely fragmented (i.e. no sub-population estimated to contain more than 1000 mature individuals)
   (b) all individuals are in a single sub-population.

D. **Population very small or restricted in the form of either of the following:**

1. Population estimated to number less than 1000 mature individuals.

2. Population is characterised by an acute restriction in its area of occupancy (typically less than 100km^2) or in the number of locations (typically less than five). Such a taxon would thus be prone to the effects of human activities (or stochastic events whose impact is increased by human activities) within a very short period of time in an unforeseeable future, and is thus capable of becoming Critically Endangered or even Extinct in a very short period.

E. **Quantitative analysis showing the probability of extinction in the wild is at least 10% within 100 years.**

Appendix IV

List of Threatened Insectivores and Tree Shrews (following IUCN, 1995)

CRITICALLY ENDANGERED (B1 and 2c)

Hylomys parvus (Dwarf gymnure)
Chimarrogale hantu (Malayan water shrew)
Chimarrogale sumatrana (Sumatra water shrew)
Sorex cansulus (Gansu shrew)
Sorex kozlovi (Kozlov's shrew)
Soriculus salenskii (Salenski's shrew)
Crocidura dhofarensis
Crocidura jenkinsii
Crocidura negrina
Suncus ater (black shrew)
Suncus mertensi
Euroscaptor parvidens
Talpa streeti (Persian mole)

ENDANGERED (B1 and 2c)

Hylomys hainanensis (Hainan gymnure)
Podogymnura aureospinula (Dinagat moonrat)
Podogymnura truei (Mindanao moonrat)
Chimarrogale phaeura (Borneo water shrew)
Sorex cylindricauda (greater stripe-backed shrew)
Sorex sadonis (Sado shrew)
Crocidura andamanensis
Crocidura beccarii
Crocidura floweri (Flower's shrew)
Crocidura grandis
Crocidura hispida (Andaman shrew)
Crocidura malayana
Crocidura mindorus
Crocidura miya (Sri Lankan long-tailed shrew)
Crocidura nicobarica (Nicobar shrew)
Crocidura orii
Crocidura paradoxura
Crocidura susiana
Feroculus feroculus (Kelaart's long-clawed shrew)
Solisorex pearsoni (Pearson's long-clawed shrew)
Suncus dayi

Suncus fellowsgordoni
Suncus zeylanicus
Uropsilus investigator
Uropsilus soricipes (Chinese shrew-mole)
Mogera etigo
Mogera tokudae (Sado mole)
Nesoscaptor uchidai (Ryukyu mole)
Tupaia longipes (Bornean tree shrew)
Tupaia nicobarica (Nicobar tree shrew)

VULNERABLE (B1 and 2c)

Mesechinus hughi (Hugh's hedgehog)
Sorex hosonoi (Azumi shrew)
Sorex leucogaster (Paramushir shrew)
Sorex sinalis (Dusky shrew)
Crocidura beatus
Crocidura canariensis
Crocidura grayi
Crocidura palawanensis
Crocidura pergrisea (pale grey shrew)
Crocidura tenuis
Crocidura zimmermanni
Suncus hosei
Suncus montanus
Desmana moschata (Russian desman)
Galemys pyrenaicus (Pyrenean desman)
Euroscaptor mizura (Japanese mountain mole)
Dendrogale melanura (Bornean smooth-tailed tree shrew)
Tupaia chrysogaster (golden-bellied tree shrew)
Tupaia palawanensis (Palawan tree shrew)
Urogale everetti (Philippine tree shrew)

VULNERABLE (D2)

Crocidura orientalis
Crocidura osorio

LOWER RISK (subcategory Near Threatened)

Hylomys sinensis
Blarinella wardi (southern short-tailed shrew)
Mogera insularis

DATA DEFICIENT

Sorex excelsus
Crocidura armenica
Crocidura minuta
Crocidura religosa (Egyptian pygmy shrew)
Crocidura shantungensis
Suncus malayanus

Appendix V

List of Members

Dr. R. David Stone
Acting Chairman, ITSES
Conservation Advisory Services
Gai Soleil
chemin des Clyettes
1261 Le Muids
SWITZERLAND

Dr. P.J. Stephenson
Secretary, ITSES
WWF Country Office
PO Box 63117
Dar es Salaam
TANZANIA

Dr. Hisashi Abe
Hokkaido University
Institute of Applied Zoology
Faculty of Agriculture
060 Sapporo
JAPAN

Mr Alain Bertrand
Laboratoire Souterrain
C.N.R.S.
F-09200 Moulis
FRANCE

Dr. Sara Churchfield
King's College
Biosphere Sciences Division
Campden Hill Road
London W8 7AH
UNITED KINGDOM

Dr. Gordon B. Corbet
Little Dumbarnie
Newburn, Upper Largo
Fife KY8 6JQ
UNITED KINGDOM

Prof. Andres Tomas L. Dans
Assistant Professor
Wildlife Biology Laboratory
Institute of Biological Sciences
College of Arts and Sciences
University of the Philippines at Los Baños
College, Laguna 4031
THE PHILIPPINES

Dr. Gilbert Dryden
Slippery Rock University
Slippery Rock, PA 16057
UNITED STATES OF AMERICA

Prof. John F. Eisenberg
Ordway Professor of Ecosystem Conservation
Florida Museum of Natural History
University of Florida
Gainesville, FL 32611
UNITED STATES OF AMERICA

Dr. Sarah B. George
Utah Museum of Natural History
University of Utah
Salt Lake City, UT 84112
UNITED STATES OF AMERICA

Dr. Edwin Gould
Curator
National Zoo
Smithsonian Institution
Washington, DC 20008
UNITED STATES OF AMERICA

Dr. Graham C. Hickman
Department of Biology
Corpus Christi State University
6300 Ocean Drive
Corpus Christi, Texas 78412
UNITED STATES OF AMERICA

Prof. Robert S. Hoffmann
Provost
Smithsonian Institution
1000 Jefferson Drive
SW/SI 120
Washington, DC 20560
UNITED STATES OF AMERICA

Dr. Rainer Hutterer
Zoologisches Forscgunginstitut und Museum Alexander Koenig
Adenauerallee 150-164
53113 Bonn 1
Germany

Dr. Gennady V. Khakhin
Deputy Head, Wildlife Health Centre
All-Russian Research Institute of Nature Conservation and Reserves
Sadki-Znamenskoye
Moscow 113628
RUSSIA

Dr. Fred W. Koontz
Curator of Mammalogy
New York Zoological Society
185th Street & Southern Blvd
Bronx, NY 10460
UNITED STATES OF AMERICA

Dr. Lim Boo Liat
Honorary Curator and Science
Malaysia Zoological Society
Dept. of Wildlife & National Park
12 Jalan Koop. Cuepacs
Taman Cuepacs, 9th mile Cheras
Kajang, Selangor 43200
MALAYSIA

Mr Nicholas Lindsay
King Khalid Wildlife Research Centre
N.C.W.C.D.
P.O. Box 61681
Riyadh 11575
SAUDI ARABIA

Dr. William J. McShea
Researcher
Conservation and Research Center
National Zoological Park
Front Royal, VA 22630
UNITED STATES OF AMERICA

Dr. Tiziano Maddalena
Institut de zoologie et d'ecologie animale
Université de Lausanne
1015 Lausanne
SWITZERLAND

Dr. Patrick A. Morris
Royal Holloway and Bedford New College
Department of Biology
Englefield Green
Surrey TW20 0EX
UNITED KINGDOM

Dr. Martin E. Nicoll
WWF Regional Office
P.O. Box 62440
Nairobi
KENYA

Dr. Jose A. Ottenwalder
Museum of Natural History
Gainesville, FL 32611
UNITED STATES OF AMERICA

Dr. Junaidi Payne
Project Director
WWF-Malaysia
W.D.T. No: 40
89400 Likas
Sabah
MALAYSIA

Ms Ana Isabela Queiroz
Instituto da Conservacao da Natureza
Rua Filipe Folque 46 - 1o
1000 Lisboa
PORTUGAL

Mrs Lynette Rajaratnam
d/a Jabatan Hidupan Liar
Tingkat 4
Bangunan Hap Seng
Lahad Datu
Sabah
MALAYSIA

Mr Felix Rakotondraparany
Curator of Small Mammals/Reptiles
Parc de Tsimbazaza
P.O. Box 4096
Antananarivo 101
MADAGASCAR

Dr. Galen B. Rathbun
US Fish and Wildlife Service
P.O. Box 70
San Simeon, CA 93452
UNITED STATES OF AMERICA

Dr. Manuel Ruedi
Museum of Vertebrate Zoology
University of California
3101 Valley Life Sciences Building
Berkeley, California 94720
UNITED STATES OF AMERICA

Dr. Duane A. Schlitter
Curator of Mammals
Carnegie Museum of Natural History
4400 Forbes Avenue
Pittsburgh, PA 15213
UNITED STATES OF AMERICA

Dr. C.G. van Zyll de Jong
RR3
North Augusta
Ontario KOG 1RO
CANADA

Prof. Peter Vogel
Institut de zoologie et d'ecologie animale
Université de Lausanne
1015 Lausanne
SWITZERLAND

Dr. Charles A. Woods
Florida Museum of Natural History
University of Florida
Gainesville, FL 32611
UNITED STATES OF AMERICA

Dr. Terry L. Yates
University of New Mexico
Department of Biology
Albuquerque, NM 87131
UNITED STATES OF AMERICA

References

Abe, H. 1967. Classification and biology of Japanese Insectivora (Mammalia) I. Studies on variation and classification. *J. Faculty of Agriculture*, Hokkaido University, Sapporo, Japan, 55: 191–265.

Abe, H. 1971. Small mammals of Central Nepal. *J. Faculty of Agriculture*, Hokkaido University, Japan, 56: 367–423.

Abe, H. 1982. Ecological distribution and faunal structure of small mammals in Central Nepal. *Mammalia*, 46: 477–503.

Abe, H. 1988. The phylogenetic relationship of Japanese moles. *Honyurui Kagaku*, 28: 63–68. [In Japanese.]

Abe, H., Shiraishi, A., and S. Arai. 1991. A new mole from Uotsuri-jima, the Ryukyu Islands. *J. Mammalogical Society of Japan*, 15: 47–60.

Bertrand, A. (1993). Répartition géographique du Desman des Pyrénées *Galemys pyrenaicus* dans les Pyrénées françaises. In: *Proceedings of the Meeting of the Pyrenean Desman*, Lisbon, Portugal. Querioz, A.I. (Ed.). pp41–52.

Biswas, B. and Ghose, R.K. 1970. Taxonomic notes on the Indian pale hedgehogs of the genus *Paraechinus* Trouessart, with descriptions of a new species and subspecies. *Mammalia*, 36: 467–477.

Blanford, W.T. 1888. *The Fauna of British India, Vol. I, Mammalia*. London.

Buckley, J. and Goldsmith, J.G. 1975. The prey of the barn owl (*Tyto alba alba*) in East Norfolk. *Mammal Review* 5: 13–16.

Butler, P.M. 1988. Phylogeny of the Insectivores. In: M.J. Benton (Ed.). *The Phylogeny and Classification of the Tetrapods, 2 (Mammals)*: 117–141. Clarendon Press, Oxford.

Cabrera, A. 1925. *Genera mammalium: Insectivora, Galaeopithecia*. Museo Nacional de Ciencias Naturales, Madrid. 232pp.

Catzeflis, F., Maddalena, T., Hellwing, S., and Vogel, P. 1985. Unexpected findings on the taxonomic status of east Mediterranean *Crocidura russula* auct. (Mammalia, Insectivora). *Zeitschrift für Säugetierkunde* 50: 185–201.

Chasen, F.N. 1940. A handbook of Malaysian mammals. A systematic list of the mammals of the Malay Peninsula, Sumatra, Borneo, Java, including the small adjacent islands. *Bulletin of the Raffles Museum (Singapore)*, 15: 1–209.

Chorazyna, H. and Kurup, G.U. 1975. Observations on the ecology and behaviour of *Anathana elliotti* in the wild. *Proc. 5th International Congress of Primatology*: 342–344.

Churchfield, S.J. 1984. An investigation of the population ecology of syntopic shrews inhabiting water cress beds. *J. Zool. Lond.*, 204: 229–240.

Churchfield, S.J. 1991. In: *The Handbook of British Mammals*. Third Edition. Corbet, G.B. and Harris, S. (Eds). Blackwell Scientific Publication.

Contoli, L. 1990. Further data about *Crocidura cossyrensis* Contoli 1989, with respect to other species of the genus in the Mediterranean. *Hystrix, New Series* 2: 53–58.

Contoli, L. and Amori, G. 1986. First record of a live *Crocidura* (Mammalia, Insectivora) from Pantelleria island, Italy. *Acta Theriologica* 31: 343–347.

Corbet, G.B. 1966. *The Terrestrial Mammals of Western Europe*. London. 264pp.

Corbet, G.B. 1978. *The Mammals of the Palaearctic Region: a taxonomic review*. British Museum (Natural History), London. 314pp.

Corbet, G.B. 1988. The family Erinaceidae: a synthesis of its taxonomy, phylogeny, ecology and zoogeography. *Mammal Rev.* 18: 117–172.

Corbet, G.B. 1992. In: *Mammals of the Indomalayan Region. A systematic review*. Corbet, G.B. and J.E. Hill (Eds). Oxford University Press, Oxford. 484pp.

Corbet, G.B. and Hill, J. E. 1986. *A World List of Mammalian Species*. British Museum (Natural History), London. 254pp.

Cranbrook, Earl of, and Medway, L. 1962. The Malayan Mole. *Malay. Nat. J.*, 16(4): 205–208.

Cranbrook, Earl of. 1966. Notes on the relationship between the burrowing capacity, size and shoulder anatomy of some eastern Asiatic moles. *J. Zool. Lond.,* 149: 65–70.

Crowcroft, W.P. 1954. *An Ecological Study of British Shrews.* D.Phil Thesis, University of Oxford.

Dans, A.T.L. 1993. Population estimates and behaviour of Palawan tree shrews, *Tupaia palawanensis* (Scandentia, Tupaiidae). Asia Life Sciences 2(2): 201–214.

Davis, D.D. 1962. Mammals of the lowland rainforest of north Borneo. *Bull. Singapore Natl. Mus.*, No. 31, 129pp.

Davison, G.W.H. 1984. New records of Peninsular Malaysian and Thai shrews. *Malay. Nat. J.* 36: 211–215.

Dene, H., Goodman, M., and Prychodko, W. 1978. An immunological examination of the systematics of Tupaioidae. *J. Mammal.* 59: 697–706.

D'Souza, F. 1974. A preliminary field report on the lesser tree shrew *Tupaia minor.* In: *Prosimian Biology.* Martin, R.D., Doyle, G.A. and Walker, A.C. (Eds). Duckworth.

Dehnel, A. 1950. Studies of the genus *Neomys* Kaup. *Annales Universitalis Mariae Curie-Sklodowska* C 5: 1–63. [In Polish: English summary.]

Dolgov, V.A. 1967. Distribution and number of Palaearctic shrews (Insectivora, Soricidae). *Zool. Zh.,* 45: 1852–1861.

Dolgov, V.A. and Hoffmann, R.S. 1977. Tibetskaya burozubka – *Sorex thibetanus* Kastchenko, 1905 (Soricidae, Mammalia). *Zool. Zh.,* 46: 1687–1692. [In Russian.]

Dolgov, V.A. and Yudin, B.S. 1975. [Progress and problems in the investigation of the insectivorous mammals of the USSR.] *Trudy Biologicheskovo Instituta, Novosibirsk,* 23: 5–40. [In Russian.]

Dötsch, C. and Koenigswald, W.V. 1978. On the reddish colouring of Soricid teeth. *Zeitschrift für Säugetierkunde* 43: 65–70.

Eisenberg, J.F. 1981. *The Mammalian Radiations: An Analysis of Trends in Evolution, Adaptation and Behaviour.* University of Chicago Press.

Ellerman, J.R. and Morrison-Scott, T.C.S. 1951. *Checklist of Palaearctic and Indian Mammals.* British Museum of Natural History, London. 66pp.

Fons, R. 1974. Le répertoire comportemental de la pachyure étrusque, *Suncus etruscus. Terre et Vie* 28: 131–157.

Fons, R. 1975. Premières données sur l'écologie de la pachyure étrusque *Suncus etruscus* (Savi, 1822) et comparison avec deux autres Crocidurinae: *Crocidura russula* (Hermann, 1780) et *C. suaveolens* (Pallas, 1811) (Insectivora, Soricidae). *Vie Milieu* 25: 315–359.

Frost, D.R., Wozencraft, W.C. and Hoffmann, R.S. 1991. Phylogenetic relationships of hedgehogs and gymnures (Mammalia: Insectivora: Erinaceidae). *Smithsonian Contributions to Zoology* 18. Smithsonian Institution Press.

Genoud, M. 1988. Energy strategies of shrews: ecological constraints and evolutionary implications. *Mammal Review* 18: 173–193.

George, S.B. 1988. Systematics, historical biogeography and evolution of the genus *Sorex. J. Mammal.* 69: 443–461.

Godfrey, G.K. 1978. The ecological distribution of shrews (*Crocidura suaveolens* and *Sorex araneus fretalis*) in Jersey. *J. Zool. Lond.,* 185: 266–270.

Godfrey, G.K. and Crowcroft, P. 1960. *The Life of the Mole* (*T. europaea* Linnaeus). Museum Press, London.

Goodman, M. 1975. Protein sequence and immunological specificity. Their role in phylogenetic studies of primates. In: *Phylogeny of Primates.* Luckett, W.P. and Szalay, F.S. Plenum Press New York. pp219–248.

Gorman, M.L. and Stone, R.D. 1990. *The Natural History of Moles.* Croom Helm, London.

Gould, E. 1969. Communication in three genera of shrews (Soricidae): *Suncus, Blarina* and *Cryptotis.* In: *Communications in Behavioural Biology* 3: 11–31. Academic Press, New York.

Gould, E. 1978. The behaviour of the moonrat, *Echinosorex gymnurus* (Erinaceidae) and the pentail shrew, *Ptilocercus lowii* (Tupaiidae), with comments on the behaviour of other Insectivora. *Zeitschrift für Tierpsychologie* 48: 1–27.

Graf, J.-D., Hausser, J., Farina, A., and Vogel, P. 1979. Confirmation du statut spécifique de *Sorex samniticus* Altobello, 1926 (Mammalia, Insectivora). *Bonner. Zoologische Beiträge,* 30: 14–21.

Gromov, I.M. and Baranova, G.I. 1981. (Eds). *Katalog Mlekopitayushchikh SSSR* [Catalogue of mammals of the USSR], Nakua, Leningrad. [In Russian.] 456pp.

Gromov, I.M., Gureev, A.A., Novikov, G.A., Sokolov, I.I., Strelkov, P.P and Chapskii, K.K. 1963. (*Mammals of the fauna of the USSR.*) 2 Volumes. Moscow.

Grzimek, B. 1975. (Ed.). *Grzimek's Animals Life Encyclopedia,* Mammals I–IV. Van Nostrand Reinhold, New York.

Guntilleke, I.A.U.N. and Guntilleke, C.V.S. 1990. Threatened woody endemics of the wet lowlands of Sri Lanka and their conservation. *Biological Conservation* 55.

Gureev, A.A. 1979. [*Fauna of the USSR, Mammals.*] Vol. 4, pt. 2. Insectivores (Mammalia, Insectivora)]. Nauka, Leningrad. [In Russian.] 501pp.

Hassinger, J.D. 1973. A survey of the mammals of Afghanistan resulting from the 1965 Street Expedition (excluding bats). *Fieldiana Zool.* 195pp.

Hausser, J. 1990. *Sorex coranatus* Millet 1982, Schabrackenspitzmaus; *Sorex granarius* Miller 1909, Iberische waldspitzmaus; *Sorex samniticus* Altobello

1926, Italienische waldspitzmaus; In: *Handbuch der Saügetiere Europas*, J. Niethammer and F. Krupp (Eds). Aula-Verlag, Wiesbaden 3/I:1.524. pp279–294.

Hausser, J., Graf, J.-D., and Meylan, A. 1975. Données nouvelles sur les *Sorex* d'Espagne et des Pyrénées (Mammalia, Insectivora). *Bulletin de la Societé Vaudoise des Sciences Naturelles* 72: 241–252.

Heaney, L.R. and Morgan, G.S. 1982. A new species of gymnure, *Podogymnura* (Mammalia: Erinaceidae) from Dinagat island, Philippines. *Proc. Biol. Soc. Washington* 95: 13–26.

Heaney, L.R., Gonzales, P.C., and Acala, A.C. 1987. An annotated checklist of the taxonomic and conservation status of land mammals in the Philippines. *Silliman Journal* 34: 32–66.

Heaney, L.R. and Rabor, D.S. 1982. Mammals of Dinagat and Siargao Islands, Philippines. *Occas. Pap. Mus. Zool. Univ. Michigan.* No. 699. 30pp.

Heaney, L. R. and Ruedi, M. 1994. A preliminary analysis of biogeography and phylogeny of *Crocidura* from the Philippines. Merritt, J. F., Kirkland, G. L. Jr, and Rose, R.K. (Eds). Special Publication Carnegie Museum of Natural History No. 18. Pittsburg.

Herter, K. 1972. Der Igel von Gran Canaria. *Bonner. Zoologische Beiträge* 18: 311–313.

Hershkovitz, P. 1977. *Living New World Monkeys (Platyrrhini)*. Volume I. University of Chicago Press.

Hill, J.E. 1960. The Robinson collection of Malaysian mammals. *Bull. Raffles. Mus.* 29: 1–112.

Hoffmann, R. S. 1984. A review of the shrew moles (genus *Uropsilus*) of China and Burma. *J. Mammal. Soc. Japan* 10: 69–80.

Hoffmann, R.S. 1985. The correct name for the Palaearctic brown, or flat-skulled, shrew is *Sorex roboratus*. *Proceedings of the Biological Society of Washington* 98: 17–28.

Hoffmann, R.S. 1986. A review of the genus *Soriculus* (Mammalia: Insectivora). *J. Bombay Natural History Society* 82: 459–481.

Hoffmann, R.S. 1987. A review of the systematics and distribution of Chinese red-toothed shrews (Mammalia: Soricinae). *Acta Theriol. Sinica* 7: 100–139.

Honacki, J.H., Kinman, K.E., and Koeppl, J.W. 1982. *Mammals species of the world: a taxonomic and geographic reference.* Allen Press, Lawrence, Kansas.

Hoogstraal, H. 1962. A brief review of contemporary land mammals of Egypt (including Sinai) I. Insectivora and Chiroptera. *J. Egyptian Pub. Health Assoc.* 37: 143–162.

Huminski, S. and Wojcik-Migala, I. 1967. Note on *Crocidura suaveolens* (Pallas 1811) from Poland. *Acta Theriologica* 12: 168–171.

Hutterer, R. 1979. Verbreitung und systematik von *Sorex minutus* Linnaeus, 1766 (Insectivora: Soricinae) in Nepal-Himalaya und angrenzenden Gebieten. *Zeitschrift für Säugetierkunde.*, 44: 65–80.

Hutterer, R. 1991. Variation and evolution of the Sicilian shrew: taxonomic conclusions and descriptions of a possibly related species from the Pleistocene of Morocco (Mammalia: Soricidae). *Bonner. Zoologische Beiträge* 42: 241–251.

Hutterer, R. 1993. Order Insectivora. In: *Mammals of the World: A Taxonomic and Geographic Reference.* Wilson, D.E. and Reeder, D.M. (Eds). Second Edition. Smithsonian Institution.

Hutterer, R. and Harrison, D.L. 1988. A new look at the shrews (Soricidae) of Arabia. *Bonner. Zoologische Beiträge* 39: 59–72.

Hutterer, R., Maddalena, T., and Molina, O.M. 1992. Origin and evolution of the endemic Canary Island shrews (Mammalia: Soricidae). *Biological Journal of the Linnanean Society*, 46: 49–58.

Ibañez, C. and Fernández, R. 1985. Systematic status of the long-eared bat *Plecotus teneriffae* Barret-Hamilton, 1907 (Chiroptera, Vespertilionidae). *Säugetierkunde Mitt.* 32: 143–149.

Illing, K. *et al.* 1981. Freilandbeobachtungen zür Lebensweise und zum Revierberhalten der ëuropaishen Wasserspitzmaus, *Neomys fodiens* (Pennant, 1771). *Bonner. Zoologische Beiträge* 27: 109–122.

Imaizumi, Y. 1960. *Coloured illustrations of the mammals of Japan.* Osaka.

Imaizumi, Y. 1961. Taxonomic status of *C. dsinezumi orii.* J. Mammal. Soc. Japan 2: 17–22.

Imaizuni, Y. 1970. *The handbook of Japanese land mammals.* Shin-Shichoa-Sha, Tokyo, 350pp.

Ingelög, T., Andersson, R. and Tjernberg, M. 1993. *Red Data Book of the Baltic Region.* Part I. List of threatened vascular plants and vertebrates. Swedish Threatened Species Unit, Uppsala.

IUCN. 1986. *IUCN Red List of Threatened Animals.* IUCN, Gland, Switzerland, and Cambridge, UK.

IUCN. 1990. *IUCN Red Data List of Threatened Animals.* IUCN, Gland, Switzerland and Cambridge, UK.

IUCN. 1991. *Protected Areas of the World. A review of National Systems. Volume 1: Indomalaya, Oceania, Australia and Antarctic.* World Conservation Monitoring Centre, Cambridge, UK.

IUCN. 1995. IUCN Red List categories. IUCN, Gland, Switzerland.

Ivanitskaya, E. Yu. and Kozlovskii, A.I. 1985. [Karyotypes of Palaearctic shrews of the subgenus *Otisorex* with comments on taxonomy and phylogeny of the group "*cinereus*".] *Zool. Zh.*, 64: 950–953. [In Russian.]

Jameson, E. W. and Jones, G.S. 1977. The soricidae of Taiwan. *Proceedings of the Biological Society of Washington* 90: 459–482.

Jenkins, P.D. 1976. Variation in Eurasian shrews of the genus *Crocidura* (Insectivora: Soricidae). *Bull. British Museum, Zool. Ser.* 30: 271–309.

Jenkins, P.D. 1982. A discussion of Malaysian and Indonesian shrews of the genus *Crocidura* (Insectivora: Soricidae). *Zool. Meded.* 56: 267–279.

Jones, G.S. and Mumford, R.E. 1971. *Chimarrogale* from Taiwan. *J. Mammal.* 52: 228–232.

Junge, J.A., Hoffmann, R.S., and Debry, R.W. 1983. Relationships within the Holarctic *Sorex arcticus-Sorex tundraensis* species complex. *Acta Theriologica* 28: 339–350.

Kaplin, A. A. 1960. *Fur-bearing animals of the USSR*. Vneshtorgizdat Publishers, Moscow. 458pp. [In Russian.]

Kawamichi, T. and Kawamichi, M. 1979. Spatial organization and territory of tree shrews (*Tupaia glis*). *Anim. Behav.* 27: 381–393.

Kawamichi, T. and Kawamichi, M. 1982. Social systems and independence of offspring in tree shrews. *Primates* 23: 189–205.

Khakhin, G.V. 1993. The conservation of [the] Russian desman. In: *Proceedings of the Meeting of the Pyrenean Desman*, Lisbon, Portugal. Querioz, A.I. (Ed.). pp79–80.

Kozlovsky, A.I. 1973. [Results of a karyological study of allopatric forms in *Sorex minutus*. *Zool. Zh.*, 52: 390–398.] [In Russian.]

Kozlovsky, A.I., Orlov, V.N., and Papko, V.S. 1972. Systematic status of Caucasian (*Talpa caucasica*) and common (*T. europaea*) moles by karyological data. *Zool. Zh.*, 51: 312–315.

Kristiannson, H. 1981. Distribution of the European hedgehog (*Erinaceus europaeus* L.) in Sweden and Finland. *Ann. Zool. Fennici* 18: 115–119.

Kuzyakin, A.P. 1965. Insectivora and Chiroptera. In: *Key to the Mammals of the USSR*, Bobrinskii, N.A., Kuznetsov, B.A., and Kuzyakin, A.P. 1965. Moscow.

Langham, N.P.E. 1982. The ecology of the common tree shrew, *Tupaia glis*, in Peninsular Malaysia. *J. Zool. Lond.*, 197: 323–344.

Lay, D.M. 1967. A study of the mammals of Iran resulting from the Street expedition of 1962–1963. *Fieldiana, Zoology* 54: 1–282.

Lekagul, B. and McNeely, J.A. 1977. *Mammals of Thailand*. Association for the Conservation of Wildlife. Bangkok. 758pp.

Lim, Boo Liat. 1967. Notes on the food habits of *Ptilocerus lowii* and *Echinosorex gymnurus* in Malaya. *J. Zool. Lond.*, 152: 375–379.

Loy, A., Dupré, E., and Stone, R.D. 1992. Biology of *Talpa romana* (Mammalia, Insectivora, Talpidae) I. Home range and activity patterns: preliminary results from a radiotelemetric study. *Rend. Fis. Acc. Lincei* 9 (3): 173–182.

Luckett, W.P. 1980. *Comparative biology and evolutionary relationships of tree shrews*. Plenum Press, New York.

Lyon, M.W. 1913. Tree shrews: an account of the mammalian family Tupaiadae. *Proc. US Natl Mus.* 45: 1–188.

MacKinnon, J. and MacKinnon, K. 1986. *Review of the Protected Areas System in the Indo-Malayan Realm*. IUCN, Gland, Switzerland and UNEP, Nairobi, Kenya.

Mahat, G. 1985. Protected areas of Bhutan. In: Thorsell, J.W. (Ed.), *Conserving Asia's Natural Heritage*. IUCN, Gland, Switzerland. pp26–29.

Malec, F. and Storch, G. 1972. Der Wanderigel, *Erinaceus algirus*, von Malta. *Säugetierk Mitt.* 20: 146–151.

Martin, R.D. 1968. Reproduction and ontogeny in tree shrews (*Tupaia belangeri*) with reference to their general behaviour and taxonomic relationships. *Zeitschrift für Tierpsychologie* 25: 409–532.

Martin, R.D. 1984. Tree Shrews. In: *The Encyclopedia of Mammals* Volume I. MacDonald, D.W. (Ed.). George Allen and Unwin, London.

McKenna, M.C. 1975. Towards a phylogenetic classification of the Mammalia. In: *Phylogeny of the Primates: a multidisciplinary approach*. Luckett, W.P. and F.S. Szalay (Eds). Plenum Press, New York.

Medway, Lord. 1977. *Mammals of Borneo*. Monogr. Malaysian Branch Roy. Asiatic Soc., no.7. 172pp.

Medway, Lord. 1978. *The wild mammals of Malaya (Peninsula Malaysia) and Singapore*. Oxford University Press, Kuala Lumpur. 128pp.

Meester, J.A.J., Rautenbach, I.L., Dippenaar, N.J., and Baker, C.M. 1986. Classification of southern African mammals. *Transvaal Museum Monograph* 5: 1–359.

Milne-Edwards, A. 1871. Descriptions of new species, in footnotes, pp92–93. In: Davis, A., Journal d'un voyage en Mongolia et en Chine fait en 1866–68. *Nouv. Arch. Mus. d'Hist. Nat. Paris* 7 (Bull.): 75–100.

Morris, P. 1983. *Hedgehogs*. Whittet Books. 124pp.

Molina, O.M. and Hutterer, R. 1989. A cryptic new species of *Crocidura* from Gran Canaria and Tenerife, Canary Islands (Mammalia: Soricidae). *Bonner. Zoologische Beiträge*, 40: 85–97.

Napier, J.R. and Napier, P.H. 1985. *The natural history of primates*. MIT Press, Cambridge.

Nicoll, M.E. and Rathbun, G.B. 1990. *African Insectivora and Elephant Shrews. An Action Plan for their Conservation*. IUCN, Gland, Switzerland.

Niethammer, J. 1964. Ein Beitrag zur Kenntris de Kleinsaüger Nordspaniens. *Zeitschrift für Säugetierkunde* 29: 193–220.

Niethammer, J. 1972. Der Igel von Teneriffa. *Bonner. Zoologische Beiträge*, 18: 307–309.

Novacek, M.J. 1986. The skull of Leptictid Insectivorans and the Higher-level Classification of Eutherian mammals. *Bulletin of the American Museum of Natural History* 183: 1–112.

Ognev, S.I. 1962. Mammals of Eastern and Northern Asia. In: *Insectivora and Chiroptera*. Israel programme for Scientific Translation. Jerusalem. 487pp.

Okhotiana, M.V. 1977. Palaearctic shrews of the subgenus *Otisorex*: biotope preference, population number, taxonomic revision and distribution history. *Acta Theriologica* 22: 191–206.

Ondrias, J.C. 1969. Die Ussuri Gross-Spitzmaus, *Crocidura lasiura*, der Agäischen Insel Lesbos. *Zeitschrift für Säugetierkunde* 34: 353–358.

Onufrienja, A.S. and Onufrienja, M.V. 1993. Desmans in the Oka State Reserve (Russia). In: *Proceedings of the Meeting of the Pyrenean Desman*, Lisbon, Portugal. Querioz, A.I. (Ed.). pp81–85.

Osborn, D.J. 1965. Hedgehogs and shrews of Turkey. *Proc. U.S. Natn. Mus.* 117: 553–566.

Palmeirin, J.M. and Hoffmann, R.S. 1983. *Galemys pyrenaicus. Mammalian Species*, No. 207. The American Society of Mammalogists.

Pavlinov, I. Ya. and Rossolimo, O. L. 1987. [*Systematics of the mammals of the USSR.*] Moscow University Press, Moscow, 282pp. [In Russian.]

Payne, J., Francis, C.M., and Phillips, K. 1985. *A field guide to the mammals of Borneo*. Sabah Society and WWF-Malaysia. 332pp.

Petter, J.-J. and Petter-Rousseaux, A. 1979. *Classification of the Prosimians*. In Doyle, G.D. and Martin, R.D. (Eds). Academic Press, New York. pp1–44.

Pitt, F. 1945. Mass movement of the water shrew *Neomys fodiens*. *Nature* 156: 247.

Poduschka, W. and Poduschka, C. 1985. Beiträge zur kenntnis der Gattung *Podygymnura* Mearns, 1905 (Insectivora: Echinosoricinae). *Sitzungsberichte der Österreichischen Akademie des Wissenschaften, Mathematisch-naturwissenschaftliche Klasse, Abteilung I*, 194: 1–22.

Price, M. 1953. The reproductive cycle of the water shrew, *Neomys fodiens bicolor* Shaw. *Proceedings of the Zoological Society of London* 123: 599–621.

Quay, W.B. 1951. Observations on mammals of the Seaward Peninsula, Alaska. *J. Mammal.* 32: 88–89.

Querioz, A.I. 1993. (Ed.) *Proceedings of the Meeting of the Pyrenean Desman*, Lisbon, Portugal.

Rabor, D. S. S. 1977. *Philippine Birds and Mammals*. University of the Philippine Press, Quezon City.

Richard, P.B. 1985. Etude préliminaire sur les rhythmes d'activité du Desman (*Galemys pyrenaicus*) en captivité (Insectivores, Talpidés). *Mammalia* 49: 317–323.

Richard, P.B. and Vallette-Viallaird, A. 1969. Le desman des Pyrénées (*Galemys pyrenaicus*): premières notes sur sa biologie. *Terre Vie* 23: 225–245.

Roberts, T.J. 1977. *Mammals of Pakistan*. Ernest Benn Ltd., London.

Robinson, H.C. and Kloss, C.B. 1918. Results of an expedition to Korinchi Peak, Sumatra. I. Mammals. *Journal of the Federation of Malay States Museum*, 8 (2): 1–81.

Robinson, H.C. and Kloss, C.B. 1922. New mammals from French Indochina and Siam. *Ann. Mag. Nat. Hist.* 9 (9): 87–89.

Rood, J. P. 1965. Observations on population structure, reproduction and moult of the Scilly shrew. *Journal of Mammalogy* 46: 426–433.

Roonwal, M.L. and Mohnot, S.M. 1977. *Primates of South Asia*. Harvard University Press, Cambridge. 421pp.

Ruedi, M. 1994. *Zoogéographie et taxonomie des crocidurinés d'asie du sud-est (Insectivora, Mammalia)*. Thèse de Doctorat, Université de Lausanne, Suisse.

Ruedi, M., Chapuisat, M., and Iskandar, D. 1994. Taxonomic status of *Hylomys parvus* and *Hylomys suillus* (Insectivora: Erinaceidae): Biochemical and Morphological Analyses. Journal of Mammalogy 75(4): 965–978.

Sàra, M., Lo Valvo, M., and Zanca, L. 1990. Insular variation in central Mediterranean *Crocidura* Wagler, 1832 (Mammalia, Soricidae). *Bolletino di Zoologia* 57: 283–293.

Shillito, J.F. 1963. Field observations on the growth, reproduction and activity of a woodland population of the common shrew *Sorex araneus* L. *Proceedings of the Zoological Society of London* 140: 99–114.

Simpson, G.G. 1945. The principles of classification and a classification of mammals. *Bull. Am. Mus. Nat. Hist.* 85: 1–350.

Sokolov, V.E. and Orlov, V.N. 1980. *Opredelitel' mlekopitayushchich Mongol'skoi Narodnoi Respubliki* [Guide to the mammals of the Mongolian People's Republic.] Nauka, Moscow, 351pp. [In Russian.]

Sokolov, V.E. and Tembotov, A.K. 1989. *Mlekopitayushchie kavkaza: Hasekomoyadnye* [Mammals of the Caucasus: Insectivores.] Nauka, Moscow. 547pp. [In Russian.]

Sorenson, M.W. 1970. Behaviour of tree shrews. *Primate Behaviour*, 1: 141–194.

Sorenson, M.W. 1974. A review of aggressive behaviour in the tree shrews. In: R.L. Holloway (Ed.). *Primate aggression, territoriality and xenophobia: a comparative perspective*. Academic Press, New York. pp13–30.

Southern, H.N. 1954. Tawny owls and their prey. *Ibis* 96: 384–408.

Spitzbergen, F. 1990. *Sorex alpinus*. In: Handbuch der Säugetiere Europas. Niethammer, J. and Krapp, F. (Eds). Aula Verlag Wiesboden.

Stewart-Smith, J. 1987. *In the shadow of Fujisan, Japan and its wildlife*. Viking/Rainbird Publication Co., London.

Stone, R.D. 1985. Home range movements of the Pyrenean desman (*Galemys pyrenaicus*) (Insectivora: Talpidae). *Z. Angew. Zool.* 72: 25–37.

Stone, R.D. 1986. *The social ecology of the European mole* (Talpa europaea L) *and the Pyrenean desman* (Galemys pyrenaicus G). Unpubl. PhD thesis University of Aberdeen.

Stone, R.D. 1987a. The social ecology of the Pyrenean desman (*Galemys pyrenaicus*) (Insectivora: Talpidae) as revealed by radiotelemetry. *J. Zool. Lond.*, 212: 117–129.

Stone, R.D. 1987b. The activity patterns of the Pyrenean desman (*Galemys pyrenaicus*) (Insectivora: Talpidae) as determined under natural conditions. *J. Zool. Lond.*, 213: 95–106.

Stone, R.D. 1989. Moles as Pests. In: *Mammals as Pests*. Putman, R.J. (Ed.). Chapman and Hall. pp65–80.

van Valen, V. 1967. New Palaeocene insectivores and insectivore classification. *Bull. Amer. Mus. Nat. Hist.* 135: 217–284.

van Zyll de Jong, C.G. 1983. *Handbook of Canadian Mammals I. Marsupials and Insectivores*. National Museum of Natural Sciences, Canada.

van Zyll de Jong, C.G. 1991. Speciation of the *Sorex cinereus* group. In: *The Biology of the Soricidae* (J.S. Findley and T.L. Yates, Eds). pp65–73. Special Publication, Museum of Southwestern Biology, 1: 1–91.

Vericaud, J.R. 1970. Estudio faunistico y biologico de los mammiferos del irineo. *Publ. Centro Pirenaico de Biol. Exp.* 4.

Vogel, P., Hutterer, R., and Sàra, M. 1989. The correct name, species diagnosis and distribution of the Sicilian shrew. *Bonner. Zoologische Beiträge*, 40: 243–248.

Vogel, P., Maddelena, T. and Sàra, M. 1992. Taxonomic status of *Crocidura cossyrensis* Contoli, 1989, with its relationship to African and European *Crocidura russula* (Mammalia, Insectivora). *Isr. J. Zool.*, 38: 424.

Walker, E.P. 1975. *Mammals of the World*. Third edition. John Hopkins University Press.

Walker, E.P. 1991. *Walker's Mammals of the World*. Nowak, R.M. (Ed.). The John Hopkins University Press, Baltimore and London.

Wang, S. 1959. Further report on the mammals of north-eastern China. *Acta Zool. Sinica* 11: 344–352.

Wang, X. 1991. *Draft review report on the course of protected areas in China*. Document prepared for the Palaearctic Asia regional Review. Third Congress on World Parks and Protected Areas, Caracas, Venezuela, February 1992.

Wang, Ying-xiang and Yang, G. 1989. [Insectivores.] pp202–210. In: [A list of medical animals in Yunnan (Yunnan Office for Endemic Disease Control, and Yunnan Sanitation and Anti-epidemic Station. (Eds).] Yunnan Science and Technology, Kunming. (not seen.)

Whitrow, G.C., Gould, E., and Rand, D. 1967. Body temperature, oxygen consumption and evaporative water loss in a primitive insectivore, the moonrat, *Echinosorex gymnurus*. *J. Mammal.* 58: 233–235.

Wilson, D.E. 1993. Order Scandentia. In: *Mammals of the World: A Taxonomic and Geographic Reference*. Wilson, D.E. and Reeder, D.M. (Eds) Second Edition. Smithsonian Institution.

Wilson, D.E. and Reeder, D.M. 1993. *Mammals of the World: A Taxonomic and Geographic Reference*. (Eds) Second Edition. Smithsonian Institution.

Wolkinger, F. and Plank, S. 1981. Dry grasslands of Europe. Council of Europe, Strassbourg. *Nature and Environment Series* No. 21. 56pp.

Wongpakdee, S. 1990. *Thailand national parks and wildlife sanctuaries*. Paper presented at the Regional Expert Consultation on Management of Protected Areas in the Asia-Pacific region. FAO Regional Office for Asia and the Pacific, Bangkok. 15pp.

Yalden, D.W. 1974. Population density in the common shrew, *Sorex araneus*. *J. Zool. Lond.*, 173: 262–4.

Yates, T.L. 1984. Insectivores, elephant shrews, tree shrews, and dermopterans. pp177–184. In: Anderson, S. and Jones, J.K. *Orders and families of recent mammals of the world*. Ronald Press, New York.

Yoshiyuki, M. 1988. Notes on Thai mammals 1. Talpidae (Insectivora). *Bulletin of the National Science Museum (Tokyo), Ser. A* 14: 215–222.

Youngman, P.M. 1975. Mammals of the Yukon Territory. *National Museum of Natural Sciences (Ottawa), Publications in Zoology* 10: 1–192.

Zaitsev, M.V. 1988. [On the nomenclature of red-toothed shrews of the genus *Sorex* in the fauna of the USSR.] *Zool. Zh.*, 67: 1878–1888. [In Russian.]

Zheng Changlin and Wang Sung. 1985. [On the insectivore fauna of Qinghai-Xizang (Chinhai-Tibetan) Plateau, China.] *Acta Theriol. Sinica*, 5: 35–50. [In Chinese; English summary.]